Presented to
**Clear Lake City - County Freeman
Branch Library**

By
Friends of the Freeman Library

Harris County
Public Library

your pathway to knowledge

DISCARD

The Last Drop

The Politics of Water

Mike Gonzalez and Marianella Yanes

PlutoPress
www.plutobooks.com

First published 2015 by Pluto Press
345 Archway Road, London N6 5AA

www.plutobooks.com

British Library Cataloguing in Publication Data
A catalogue record for this book is available from the British Library

ISBN   978 0 7453 3492 9   Hardback
ISBN   978 0 7453 3491 2   Paperback
ISBN   978 1 7837 1520 6   PDF eBook
ISBN   978 1 7837 1522 0   Kindle eBook
ISBN   978 1 7837 1521 3   EPUB eBook

This book is printed on paper suitable for recycling and made from
fully managed and sustained forest sources. Logging, pulping and
manufacturing processes are expected to conform to the environmental
standards of the country of origin.

Typeset by Stanford DTP Services, Northampton, England
Text design by Melanie Patrick
Simultaneously printed by CPI Antony Rowe, Chippenham, UK
and Edwards Bros in the United States of America

Water is sometimes sharp and sometimes strong, sometimes acid and sometimes bitter, sometimes sweet and sometimes thick or thin, sometimes it is seen bringing hurt or pestilence, sometimes health-giving, sometimes poisonous. It suffers change into as many natures as are the different places through which it passes [...] it is warm and is cold, carries away or sets down, hollows out or builds up, tears or establishes, fills or empties, raises itself or burrows down, speeds or is still [...] In time and with water, everything changes.

Leonardo da Vinci

Contents

List of Figures, Table and Boxes

Preface
The Road to Machachi

In the distance, an eternal snow melting on its slopes, was Cotopaxi, the volcano whose name in Quichua means 'the soft throat of the moon'. Poetic and menacing, it was ahead of us all along the road to Machachi. It still wasn't clear to us where we were going. We were travelling with various indigenous organisations, moved by that concept of 'good living' that we still hadn't fully absorbed, to a land occupation.

What the comrades from ECUARUNARI and CONAIE, Ecuador's indigenous federations, were describing to us was an everyday activity for them, but we were uncertain what to expect. As the days of the revolution slipped by, these people were still condemned to poverty. And they pursued their unfulfilled demands with machetes hidden in long skirts and steel in the soul.

Figure 0.1 Preparing for a land occupation, Machachi, Ecuador
© Mike Gonzalez

They gathered on open ground then climbed into a line of open lorries; it felt like a prelude to conflict. It wasn't far to the land they were about to occupy, land taken by a man called Jaramillo from its original inhabitants. In the beautiful little town square, at its centre a tableau of men and women cultivating the land, the women waited under the trees, their children beside them. The men gathered to confront the landowner. Around the square the wind blew dust from the earth streets into the air. Some of the women who lived in the town looked out and their children came to point and laugh and ask who we were. Did they have water in their houses? No, was always the answer. It was in barrels that stained the water brown. How did they manage with cooking, washing, bathing the kids? Mr Jaramillo opens a standpipe every evening and lets us take some water. And at Christmas he leaves it open so we can fill the barrels and he gives us a pig and a sack of potatoes each. He's a good man. But didn't he take the land by force? 'Some people say that, but he lets us live here in peace. There's no point protesting. He'll never go.' A young woman with her kids smiled and gave me a bottle of Orangina. It was welcome. It was midday and the heat was rising.

People rushed towards a corner of the square. The landowner and his daughter, illegitimate occupants of this fertile land they had fenced in and called their own, had arrived. The spokespeople for ECUARUNARI showed him the document that gave them the right to take back their land. Jaramillo waved his title deeds with their seals and signatures. The National Guard and a representative of the mayor arrived to prevent a violent confrontation. Jaramillo insisted 'my lands are productive, we work hard.' The man from ECUARUNARI said 'you grow flowers; you can't eat flowers.' The crowd pushed towards him, all speaking at once, aggressive and challenging. The woman from the mayor's office couldn't make herself heard above the shouts from both sides. I moved away from the circle. A man stood looking towards the crowd, his face tense with rage. He was one of Jaramillo's workers.

He'll never leave Machachi, because he owns the water.

The water? Isn't it just the land he owns?

He looked back at me. The water is in the land. The spring is up there and he controls it.

To water his flowers?

For his flowers and for Orangina. He sends the water to the bottling plant. The tankers come every day to collect it.

And the people here? Aren't they thirsty?

At that moment a guard came towards us. The man stopped. 'If I say any more I'll lose my job,' he said, and walked away, still thirsty and still angry. I moved to follow him but the crowd moved in on Jaramillo, his daughter and the mayor.

The guards formed up in front of the gate that Jaramillo to keep the community away from his empire. 'This land is my father's. That's a lie; it says here that he has to give it back to its original population, to grow food. I'm mayor here; now calm down and listen to me. When the court order gets here you can go.'

How long will that take?

They said we'd have it tomorrow.

In one voice they shouted back. 'Then we will stay here and wait until it gets here!'

Acknowledgements

This is for David MacLennan, for many reasons, and for our children Dominic, Anna, Rachel, Ezequiel, Eleazar and Ernesto and our grandchildren Amara, Max, Luca and Martha Lucia, for putting up with us. We would like to thank the many people who in one way or another have helped to make this book possible, especially Carlos Zambrano of Camaren in Quito, the comrades of CONAIE and ECUARUNARI in Ecuador, the community of Tzarwata in the Rio Napo in the Ecuadorean Amazon, Wilmer who risked his safety to show us what was happening in Yasuni, Oscar Olivera and Marcela Olivera of the Coordinating Committee for Water and Life, Cochabamba, Tom Lewis, Richard Boyd-Barrett, Paul O'Brien and Brid Smith of the Right2Water Campaign in Ireland, Marcelle Dawson who shared her work on the South African Movement so generously, Trevor Ngwane for his inspiration, Pueblo Nasa, La Red Latinoamericana de Mujeres, No a la Mina, Icíar Bollaín and Paul Laverty for their film *Even the Rain*, Manuela Blanco for her film *El río que nos atraviesa*, Alex Bell, Alex McCall and June Toner, with an apology, Frank Poulsen, Peter Archard and Ayesha Usmani, Jonathan Neale, Pat Collins and Magdalena Villalobos, Scott Donaldson, Claudia Abache and Carlos Rivodó for their solidarity, Pepijn Brandon and Willemijn Wilgenhof for their friendship and insights, Paddy Cunneen for nearly everything, and David Castle of Pluto Press for his confidence in us and his quiet encouragement.

Introduction

Where there is water, there is life. Without it, there is nothing. Our collective nightmares about the future always seem to centre on worlds turned into deserts. Some 85 per cent of the planet's surface is water. And much of what is not, from the land to the people who inhabit it, is also largely water. It sustains the body, nourishes the land, drives the wheels of industry, and transforms itself in many unexpected ways.

So why is it that in recent decades, the talk of a water crisis has risen to a cacophony, when for so much of our history we have assumed its availability and its continuing flow? How can there be a shortage of something that is everywhere we look, and that regularly cascades over us. Why has the talk of crisis suddenly become so insistent?

The reality is that there is not one crisis but several, however, according to the World Water Report of 2014, 'the crisis is essentially a crisis of governance'. In other words, it is not a natural phenomenon we are discussing, not simply that there is so much water and the world's population is growing. The problem is one of the management and allocation of the water that exists on earth.

It is clear that climate change is happening; it is by now part of everyone's day-to-day experience. The rainfall patterns over centuries have changed, the seas have warmed and the ice caps and glaciers are shrinking. Most Europeans know that winters are colder and the summers growing hotter. Recent tragedies have brought home in dramatic ways the changes that are to come. Hurricane Katrina has left us with tragic and terrible images of not only the impact of the hurricane, but also more importantly of the cold cynicism of people in power, who abandoned the poor of New Orleans to their fate with so little scruples. The tsunami of 2004 killed hundreds of thousands in hours, and drove home both the unpredictability and the power of water. Yet for a long time after a group of concerned climate scientists from all over the world formed the IPCC (the Intergovernmental Panel on Climate Change) in 1989, there was an often vicious debate calling their carefully researched and reported conclusions into doubt.

The 'climate sceptics' who publicly questioned them were supported and financed by 'Big Oil' and 'Big Coal', whose influence and enormous wealth stemmed from industries that, according to the scientists, were significantly responsible for emitting the greenhouse gases that were leading to global warming and climate change.

The decade of 2005–2015 was declared to be the Decade of Action for Water and Life; the decade when the optimistic Millennium Goals would be fulfilled, among them the reduction by half of the 1.2 billion human beings[1] still without access to clean drinking water as the century began, and the 2 billion plus without sanitation (a figure much less quoted but in many ways far more significant).[2] The international community's decision to support these goals stemmed perhaps from the apocalyptic pronouncements that intensified as the twentieth century drew to its end, and perhaps most dramatically the much repeated prediction by a deputy director of the World Bank that where the twentieth had been a century of conflicts over oil, the twenty-first would be marked by water wars. The wreckage left by the competition for oil surrounds us still, from the systematic destruction of Iraq and Afghanistan, to the endless crises in the Middle East. There seem not yet to have been similar acts of wanton destruction in relation to water. Was the prediction wrong? Not at all, but there are many ways to wreak havoc, and many levels on which the water conflict will express itself. In fact, there are struggles everywhere over water, as significant and dangerous as the wars of position conducted between unarmed drones and rifles among ruins that we have become so accustomed to. The water wars, if they are allowed to happen, will be conflicts between desperate people facing 'a general degradation of their living standards'.[3] The struggles are beginning, and they are not restricted to the poor regions of the world; there are mobilisations everywhere as the awareness of the importance of water grows. As we write, Ireland is living through a determined protest movement against water privatisation, great marches cross India again and again to demand democratic control of water, China faces water scarcity at a level so dramatic that massive tunnels are being built to carry water back to its northern regions, while in Brazil 100,000 indigenous families are being displaced by the Belo Monte dam which will flood their home territories to feed local metallurgical industries. There are insistent predictions of recurring drought in the United States and the implications of the shrinking ice cap for the planet's future are part of the daily diet of public anxieties.

For the most part, however, these conflicts are of a different order, where the opposing forces are deeply unequal. The rich world of the north is prolific in its use of water, and a good proportion of the world's fresh water is located in these regions; the water-poor live in the developing world where the public provision of water is a late development, and often unsystematic and subject to powerful external pressures. The increasingly common picture is, on the one hand, the control of water supplies by the huge multinational companies grown confident and powerful in a neo-liberal era, and on the other public enterprises under siege and local communities whose only weapons are resolution, mass organization and the combination of conviction and, in many cases, despair. The iconic struggle of the communities of Cochabamba in Bolivia in 2000[4] against just such an enemy, the Bechtel Corporation, was the first concrete demonstration that even such unequal battles could be won.

As to the Millennium promises, they remain largely unfulfilled; indeed the battle to provide water to the poorest on earth – and they are not only to be found in desert regions or the communities of Central Africa – still has far to go. The *UN Water Development Report 2015* presents figures for progress towards the Millennium Development Goals (MDG) targets that show 69 countries 'seriously off target' – in other words unlikely to reach the sanitation targets by 2030 – and 53 countries that will not meet the objectives for access to safe drinking water.

Conflicts over water are about far more than the provision of water in itself. Water, after all, is part of that range of resources that should belong to all – the *commons* – that have been appropriated and privatised at an increasing pace, as their distribution too has grown increasingly unequal. The response from the global market has been to attack, directly and indirectly, the idea that the earth is 'a common treasury for everyone to share',[5] and redefine water as 'an economic good' as opposed to a human right. In the slippery vocabulary of neo-liberalism that means that it should only be available to those who can pay. And what of the rest of humanity? What of those who were to be the beneficiaries of the Millennium goals conceived in an uncharacteristic moment of sentimental concern by the world's ruling classes as they faced the imminent collapse of the Millennium clock? They were easily forgotten in the brutal realism of the age of globalisation.

The explosion of writings about water[6] reflect a real change in its use and distribution from the last two decades of the twentieth century onwards.

If for most of human history rain and rivers served human needs, in the industrial age rivers and lakes were sourced both for water for human use but also to drive the turbines of industry. The difficulty is that the discussion has centred on drinking water and, very much second, on sanitation and by extension on the individual and domestic uses of water. Perhaps this is the level at which we can grasp the significance of water in our lives. But the water use that is changing the face of the planet, with unanticipated and often hidden consequences, is not all visible to the naked eye. While just 10 per cent of the planet's water is dedicated to domestic consumption, the rest is divided between industry and agriculture, with the lion's share (around 65%) going to agriculture. The population of the planet is increasing and with it food production. But agricultural production embraces much more than food crops – maize and sugar are harvested to produce bio-ethanol, an alternative fuel for cars and machines, for example. But it is neither cheaper nor environmentally friendlier than the fossil fuels on which we have come to depend so heavily. Oil production involves huge quantities of water, and increasing amounts as other reserves are mobilised – tar sands, shale gas and fracking, for instance – which require even more water in the extraction process. But the decision to develop ethanol was not based on any considerations about the best use of water, or any other resource, but in anticipation of 'peak oil' – the point at which half of the total oil available on earth had been exhausted – a point which may already have been reached but which is anyway close. The arrival of that critical moment could have generated serious debates about rationalising, controlling and reducing our palpably wasteful use of oil – millions of private cars circulating with a single passenger while public transport is savaged everywhere, for example – as well as the huge proportion of other apparently unrelated goods which are by-products of oil – plastic in its million and one manifestations, nylon and many others.

In the late 1960s, as population growth became an issue of public debate, and amid dire predictions that current food production could not keep pace, a 'Green Revolution' was announced that would multiply the productivity of food crops. Not for the last time, the word 'green' laid a reassuring cover over large-scale genetic modification and the use of chemicals and pesticides, but also concealed a massive increase in water use in agriculture. This was not addressed at the time, nor for many years thereafter, because water was regarded as a virtually infinite resource. Rivers were there to be seen, the exaggerated optimism of a dam-building age gave us huge man-made lakes. Industrialisation spread into the developing world, and its cities grew at an accelerated speed, partly at least as dams and an industrialised agriculture expelled growing numbers from the countryside. By the later 1980s it was also possible to speak of water *mining* on a large and growing scale. The world's rivers and lakes were insufficient to respond to society's water needs, and the underground aquifers that hold 30 per cent of all freshwater began to be mined – wells sunk deeper and deeper and human ingenuity placed at the service of finding ways to draw the underground water at a faster and faster rate, passing many times over the rate of recharge. Aquifers are recharged by the process whereby precipitation filters through or *infiltrates* the soil to replenish the aquifers, or flows into rivers and lakes as *run off* before they in their turn flow into the sea whose waters will evaporate at varying rates to become water vapour and in turn the rain that sustains the water cycle. A third source is *evapotranspiration* – the water caught within living beings that adds to the water in the atmosphere.[7] This water cycle was the guarantee of the survival of the species – for we are ourselves 70 per cent water. Harnessing water has been a permanent feature of human societies – storing it and diverting it to the crops in the fields, to sustain life. The Roman engineers have left evidence of a new phase, the transporting of water, wherever their aqueducts are framed against the sky. Other ancient civilisations – Babylon, Athens, Ur – developed sophisticated means of transporting water too, controlling and to some extent taming nature. And as Wittfogel famously analysed, in such 'hydraulic societies' the control of water also brought with it power and inequality. But it is only with the industrial age that the power of water begins to be harnessed for commercial gain.

> During the first three-quarters of the 20th century absolute and per capita demand for water increased throughout the world. Freshwater withdrawals increased from an estimated 500 cubic kilometres per year in 1900 to 3580 cubic kilometres per year in 1990.[8]

The calculation per capita does not provide us with a true picture. In some countries, the USA and Canada for example, individual water use is far above the world average; in other parts of the world, in particular Africa and the Middle East it is far below that per capita figure. But beyond the unequal distribution of water, a critical issue as we address the likely shortfall of the Millennium Goals by 2025, the reality is that industry and agriculture are not only the direct users of the bulk of water, but they are also responsible for the shrinking availability of water overall. Industrial processes not only use water; they pollute and contaminate it, so that it will not and cannot be returned, but will reduce the overall amount of available water year on year.

The most significant change in the late twentieth century, however, was that the increasingly intense public discussion about water scarcity produced not only a debate about how best to use our water, what processes we could develop with our extraordinary technological ingenuity, what level of wastefulness characterised our production systems, but rather a new enthusiasm among those who controlled the global capitalist economy. A new and potentially immensely profitable commodity had appeared in the global market place – water. Bottled, channelled, transported towards the richer corners of the world, it could yield enormous profits. How ironic, then, that the possibility of water scarcity, or rather its likelihood, could be seen as an *opportunity* rather than a problem it was incumbent on the whole of humanity to resolve!

Our starting point is that water is a key component in the maintenance of a decent human existence. That much is obvious, though the fact that it has to be argued at all exposes the cruelty of the global system. Once it was sufficient to dip a hand in a nearby lake or river, or channel its flow to irrigate the crops. Wittfogel's definition of 'hydraulic societies' described how human civilisations have arisen (and fallen) around water through thousands of years. But those water sources have become contaminated, polluted by the poisons of human invention, or simply drained, their natural cycle of renewal interrupted and undermined. And that has happened not only because civilisations and their cities have grown well beyond the river banks, nor simply because industrialisation has accelerated the rhythm of change, but because it has happened in the framework of capitalism. Production may seem to be driven by techniques and machinery, but they in their turn are driven by values, purposes and the yearning for accumulation of those who own them. In that system of values water becomes a commodity,[9] its use and allocation determined only by its market price. The sometimes abstruse arguments about use and exchange value become suddenly very clear when the subject is water. Water is life itself, as the cliché repeats in almost every language; yet today it is subject to the laws of the market, and to its contempt for life. When neo-liberal theorists describe water as an 'economic good', it is placed in the same category as an automobile or a dress from Dior. It is available only to those who pay, and those who cannot will suffer the predictable consequences. It is not a question here of who should pay less or more. To ask it about water at all is to come face to face with the central contradiction of capitalism. Water is for the benefit of all, a common resource. How to ensure that it remains so is our central concern as authors and activists. But just as water flows into every crevice so its democratic control affects every single human activity; a new and just world water order is only imaginable in a world governed by different values and shared collective purposes. Just as most water wars have in fact been local confrontations, so a new world water order will begin with local collaborations multiplying on a global scale. But it is an urgent matter, as we will show, to ensure that we never have to contemplate the fate of the last drop.

1

A Floating Planet

The failure to provide drinking water and adequate sanitation services to all people is perhaps the greatest development failure of the 20th century.[1]

The presence of water

Of all the water on earth, 97.5 per cent is too salty to drink. Two-thirds of the remaining 2.5 per cent is locked in the ice caps at the poles or in snow, though these areas are shrinking with the impact of global warming. There is water in the air – water vapour and then rain – and there is water held inside plants and other living beings, like ourselves. That leaves 16 million cubic kilometres, much of which is trapped in sedimentary rock too far underground to access. A further 90,000 cubic kilometres is in rivers and lakes. 500,000 cubic kilometres per year evaporate from the sea and from living plants, though about 60 per cent of that is returned in the form of snow or rain.[2] The water held underground in aquifers, *groundwater*, has taken millions of years to accumulate. So the main source of fresh water, other than aquifers, is 'run-off', the water that seeps through the soil into rivers and lakes from their banks. This represents something like 34,000 cubic kilometres,[3] which is about twice what is currently used. Rainfall amounts to some 110,000 cubic kilometres of which half is trapped by vegetation and around 30 per cent falls into rivers and lakes.[4]

The problem is that the gross figures for rainfall and river flows do not show how unevenly distributed that rainfall is, nor the differing rates at which aquifers, lakes and rivers assimilate the run-off. The flow from river bank into rivers is visibly rapid, though the speed of flow will be affected by the condition of the river banks, the amount of vegetation and the number of trees. The recharging of water into the shallower aquifers through the soil takes longer, and is not to be hurried, but penetration into the deeper water basins can take centuries, or never happen at all. Fred Pearce estimates that some 14,000 cubic kilometres are really accessible. Rivers, after all, flow through networks which when seen from the air resemble the jumble of veins in our hand; they are systems rather than single courses flowing neatly down a predetermined route. Indeed if the meandering of a river is disturbed, or the course of the river is straightened, and shortened, swamps will be drained and the run off will rush to the sea more quickly instead of penetrating the soil. In some seasons of the year their flow turns to flood, expanding across flood plains; at others it dries up. Rivers are by definition anarchic,[5] and a great deal of human effort and resourcefulness over time has gone into attempting to control or direct their flow to guarantee stable supplies of water, especially for irrigation. One result has been the proliferation of dams across the world, built in the optimistic expectation that they would guarantee water supply and energy provision for generations; their unforeseen consequences are now emerging in the form of huge displaced populations, the accumulation of silt which would previously have flowed with the river and dispersed on deltas and flood plains, and which reduces the amount of water in reservoirs, the spread of poisonous algae as a result of inadequate oxygenation of the water, and the stagnant water and rotting materials expelling methane gas into the air from the surface of the reservoirs, among others. In the USA these monuments are now being rapidly dismantled; elsewhere, and particularly in China, they continue to be built on an increasingly massive scale, with accelerating negative consequences.

And the problem is even more complex than that. Humanity has controlled and contained rivers – it has also polluted them. The refusal of rivers to acknowledge national frontiers has produced other kinds of conflict, to which we shall return. The rain, while replenishing the planet's resources, falls unevenly – flooding here, while leaving deserts there. And the supply of water from rivers, even after the construction of dams, has proved inadequate for the needs of industry, agriculture and people, so attention has turned to the giant aquifers beneath the ground. The current statistics for the depletion of aquifers are truly alarming, especially in the knowledge that the deeper water basins are pretty much inaccessible.

It is easy to be blinded, or depressed, or confused (or all three) by statistics, especially with these astronomical numbers. What we need to know is how much water is available to us, where that water is, how it is currently used, and how we can use it differently. It is more than an issue of conservation – though there is clearly an urgent need to husband, control and conserve our water supplies over time. Recent studies[6] have begun to speak more and more insistently about 'peak water', drawing an analogy with the argument about 'peak oil'. There is an undoubtedly catastrophist element to these discussions. We know that fossil fuels do have a tangible limit, although there are immense reserves still to be exploited – in Venezuela, Iran, Bolivia, Saudi Arabia for example; but the reality is that oil cannot be regenerated and that its available quantities are finite. Fracking has opened a new avenue of supply, but with consequences for the stability of the planet about which we still know very little, and with disastrous consequences for the water table which remain to be calculated, although a recent report in the United States indicates that the water used in the extraction process shows a level of benzene contamination that is 'off the charts'.[7] The analogy between oil and water, however, does not apply, for one very simple reason. Water is a renewable resource – though with humanity's sometimes diabolical ingenuity, it is probably possible that it could be contaminated to such a degree that large numbers of people could be denied the most essential component of survival. One prediction has it, for example, that two-thirds of world's population will live in water-stressed conditions by 2025, *if current consumption continues*.[8] But that will occur only to the extent that we allow the current system of production, with its uncontrolled and unregulated use of water, to continue. 'Peak water' is not the consequence of limited supply but of a specific pattern of use – and that is what has to be urgently addressed.

Six countries have between them half of the world's renewable fresh water: Brazil, Canada, Russia, Indonesia, China and Colombia. Some 14,000 cubic kilometres of water are in rivers; the Amazon, the Congo and the Orinoco hold 25 per cent of the total, but none of them are in areas that are easily accessible. River water that can easily be reached and used probably amounts to around 9,000 cubic kilometres. Boiled down to individual consumption that would represent about 1,400 cubic metres per person per year, though a typical individual consumer in the West could use between 1,500 and 2,000.[9]

Inevitably, the mention of consumption conjures up images of individuals and households using their water in tangible ways – to flush toilets, water lawns, wash clothes or dishes, clean the car and so on. The emphasis is on individual and household consumption. And there are certainly many ways in which fresh water can be conserved on quite a large scale – by recycling waste water, for example, or reconsidering the symbolic cultural value of lawns in dry areas (like California), severely limiting the use of garden hoses and even – that holy of holies – restricting the number of golf courses that are built, for golf courses are among the most demanding consumers of water on behalf of a tiny clientele. These are some of the many tangible measures which will improve the way we use water, reduce quite significantly the impact of overuse and misuse of water on the environment and raise the level of awareness of people in this and other areas. But domestic uses of water account for only 10 per cent of the total; 25 per cent is currently consumed by industry and the remaining 65 per cent by agriculture.

The situation is further complicated by the issue of *distribution* of water resources; water scarcity – and it is important to emphasise, as Maggie Black insists we do,[10] that it is not only water to drink that we are discussing, but water to wash in, water to flush away human detritus – affects the world's population in very distorted ways. The 2.6 billion people without adequate sanitation are concentrated almost entirely in the developing world. There are poor communities in Europe and North America, of course, but water continues to be available to them – for the present. This picture will surely change, however, if it becomes generally acceptable to view water as a consumer good, which people can have to the extent that they can pay for it.

For most of human history, it was rivers and lakes that provided for the water needs of human societies. The rise and decline of civilisations, the 'hydraulic societies', were directly attributable to the management (or mismanagement of water). But with industrialisation and the growth of cities, especially in the twentieth century, rivers were dammed to create artificial reservoirs – or man-made lakes – where natural watercourses could not provide for growing populations increasingly concentrated in urban spaces. As the dams grew in size and the reservoirs in capacity, the water cycle – that process of renewal and replenishment of water that ensures that water is a renewable resource – was progressively disrupted, with long-term effects that were only recognised at a much later stage.

The Colorado, on which the Hoover Dam stands, once reached Mexico and flowed into an estuary 40 kilometres wide. Today, it does not reach the delta at all. The Indus dries up and disappears 80 miles from the sea, while China's giant Yellow River failed to reach its estuary for nearly eight months a year in the 1990s and today barely reaches it at all. The course of many rivers has been altered, rapidly drying out fertile wetlands, just as the flood plains once reached by dammed rivers – fertile places for a seasonal agriculture – have now shrunk or disappeared. The emblematic dam of the post-Second World War era is the Aswan, the great dam on the Nile in Egypt eventually financed by Russia, after the USA withdrew its support from the increasingly independent and nationalist Nasser government. Work on it began in 1960 and the High Aswan Dam was completed ten years later. The Nile originates in Sudan and flows through Ethiopia to Egypt, which takes 77 per cent of its flow – the lion's share. The giant Aswan Dam blocked the river, interrupting its flow towards the flood plain of the Nile delta. The bulk of its water is used to irrigate the Upper Nile, where cotton is the principal crop. Yet Egypt is a major importer of wheat – though the inadequacy of supplies was illustrated in Egypt's bread riots in 2004.

It is now generally recognised that Lake Nasser, the reservoir above the High Dam, loses 15 per cent of its volume annually in evaporation, releasing methane gas. And like all dams, the build-up of silt in the lake bed – silt which would previously have swept down towards the delta, de-oxygenates the water, where poisonous algae can now proliferate, and over time reduces the lake's capacity. This is not exclusive to Lake Nasser, where large budgets have gone into digging drainage channels to prevent the accumulations.

There are unlimited examples of dams and diversions drying the lower reaches of otherwise healthy rivers. The most notorious example is the Aral Sea, in what was Soviet Central Asia. It is the 'ultimate ecological cautionary tale'.[11] Once the fourth largest inland sea on the planet, two major rivers flowed into the Aral, the Amu and the Syr, originating in the Himalayas. In the 1950s, Soviet engineers diverted the rivers to grow cotton in what was a desert terrain. The waters of the Sea receded, leaving what were once islands at an increasing distance from the sea. The fishing industry, on which 40,000 local families depended, was decimated as the concentration of salt killed the fish. Wildlife disappeared and irrigation increased the levels of salt, which then melded with pesticides to create what Maggie Black describes as 'a stew of poisonous sediments'. The climate changed, making it impossible to grow cotton after all; farmers switched to rice, which consumed even more water. After the collapse of the Soviet Union, five Central Asian states found themselves jointly responsible for the Sea. To date nothing has been done and all that remains is a grotesque landscape of boats stranded in a desert as camels make their way past them.

Far less known, but in many ways even more catastrophic, is the disappearance of the Hamoun wetlands on the Iran–Afghanistan border.[12] Before the discovery of oil and gas in the region, the Hamoun lakes covered 5,600 square kilometres serving nearly half a million people in 935 villages. But in the 1990s, Afghanistan's control of the Helmand River and the Kajaki Dam, built by US engineers, spelled disaster for Iran. In 1998, the Taliban effectively cut off the two Hamoun lakes in Iranian territory until the Afghan reservoir filled – which it rarely did. The years of drought that followed, added to the failure of the irrigation works at the dam, and transformed the region into the dry place it is today.[13] Now, like the Aral Sea, all that remains is parched dry land where derelict boats lie rotting. The local populations have suffered raised rates of heart disease and respiratory disorders are general, the result of the contaminated dust that floats up from the old lake bed. In Africa, Lake Chad, bordered by four African nations – Niger, Nigeria, Cameroon and Chad – was once 25,000 square kilometres of inland lake that supported two million people from fishing and farming on the lake shores. It was once again a story of drought combined with the diversion of water in ill-conceived irrigation projects that reduced the great lake to 2,500 square kilometres, and its fishing communities to farmers cultivating local crops – sorghum, millet and cowpea; when rice was introduced, and dykes were built around the paddies, the river was deprived of part of its flow and the flood plain ecosystem was undermined. In China, meanwhile, the unregulated and breakneck development of industry and industrial agriculture has polluted 80 per cent of its rivers, almost all of which flow down from the Himalayas through China to India, Pakistan, Bangladesh and Thailand.

The story these examples tell is in one sense very complex, but in another sense can be summed up as the consequence of ill-considered water projects imposed largely for financial reasons by international agencies and aid programmes, the bulk of which were designed to profit industries and investors in the donor countries. Thus, for example, external insistence on providing aid for the digging of deep tubewells in the Indus River Basin took no account of the naturally occurring arsenic in the aquifers of which millions of people have been or will become victims.

Water projects, if they are to be what they claim to be – projects for development rather than profit for capital – should be prepared and thought through in collaboration with the people who are most intimately familiar with each region, the local farmers and communities who know how to read their own land and their own climate. But because development rarely means social transformation for the benefit of the population, schemes are imposed using the wrong criteria – with catastrophic effects.

The hole in my bucket

The statistics tell us that there is more than enough water on earth to supply the needs of the world's population. And yet there is an accelerating water crisis that is affecting above all the world's poorer communities. So where is the water going?

The rise of industrial society and the growth of cities for the first time separated human beings from their water sources. Growing populations crowded together in the slums and shanty towns of the nineteenth-century city. The diseases that proliferated there were not at first linked to water; the prevailing doctrine, the miasma theory, held that disease was spread through the air. But major cholera outbreaks and the pioneering work of mid-nineteenth-century visionary scientists like Edwin Chadwick, author of the historic 1842 *Report on the sanitary conditions of the labouring population of Great Britain*; Lemuel Shattuck who, under the influence of Chadwick, wrote the first comprehensive public health plan in the United States; or Dr John Snow in London, who in 1854 established the link between the ravages of cholera and water, and initiated the discussion of water and public health. Snow did not, as yet, understand the relationship between cholera and contaminated water, and nor did the British parliament until the Great Stink, when the intolerable smell of a fetid Thames forced them into action. In Glasgow too, it was a cholera outbreak that led to the construction of a pipeline from Loch Katrine into the city in 1859. The result was a doctrine of public health that in the succeeding four decades or so led to serious municipal commitments to the provision of clean water and sanitation. The solution was seen to lie with engineers rather than politicians, and the costs would be borne by the public purse. There was an essentially humanistic view underpinning many of the discussions of public health, heavily

Figure 1.1   In Africa water will often be a long walk from home. © Carlos Rivodo

underpinned with Victorian moralism and a dose of paternalism. There was also, of course, a barely hidden financial motivation – sick workers were not productive workers. Critically, these pioneering projects established a link between the provision of drinking water and systems of sanitation. This connection has been central to water management in Western Europe and North America from the outset. In the current debates, however, the two have become separated – the argument about drinking water is more easily understood and less uncomfortable than long discussions about human waste. More recently, the international debate has turned to sanitation, especially after the major outbreak of that most emblematic of diseases – cholera – in Peru in 1990, and in KwaZulu, South Africa, in 2000.

In nineteenth-century Europe, water provision was largely a municipal responsibility, paid for from local taxes and locally run. The sources were lakes and rivers in the vicinity, and the transporting of water over distances was not a major issue. What was critical was adequate engineering, both to carry the water through pipes and to control its flow and direction. The key was planning. 'Until very recently [the] dominant ideology was not a financial requirement to cover costs nor an ecological need to conserve resources but the engineering priority of meeting growing demand.'[14]

This was not quite as selfless as it sounds. The beneficiaries in the larger scheme of things were also the employers, who could count on an increasingly healthy workforce. And the costs of these engineering projects would be borne out of the profits of industry. Urban planning made it possible to lay pipes and sewers to service households as well as factories. This public health model served the industrial cities as they grew, and was adequately sustained and organised by municipal water authorities.

But water was more than simply to drink and bathe in; it was also energy, driving turbines and producing electricity where coal was not available. At an early stage the water wheel was emblematic of this area of water use; later the Hoover Dam became the icon of the control of water resources to generate power in a second industrial age. And it symbolises the criteria which have increasingly prevailed in the organisation of water resources and in the public debates around them. In the rush to provide water for an expanding intensive agriculture and a burgeoning industry, public provision was left behind. Municipal water providers, subject to the limitations of local budgets and the seductions of small-scale power broking and corruption, did become increasingly inefficient. The money available for the large-scale maintenance of urban pipes and pumps shrank. For anyone growing up in a Western European city leaks were a fact of daily life and rarely considered; water, after all, would always be there. But this was no longer clean spring or river water. The reality was that it had to be cleaned, filtered, and treated – and this was a process that required guaranteed funding. This has become an argument repeatedly called upon to justify paying for water, abandoning without comment the conception of meeting social needs out of taxation.

And behind it lies a deeper ideological change – abandoning the concept of water as a public good.

Despite the way in which water provision is still generally discussed today – in terms of drinking water and domestic sanitation – individuals and households remain a small proportion of consumers of water. And while the world's population has certainly expanded more or less exponentially into the twentieth century and after, it was agriculture (consuming 70%) and industry (20% and rising) that were, and still are, the majority consumers by far of water; it is their needs that prevail in the organisation of water supplies. Irrigation uses the bulk of accessible river waters; industry uses water as energy in all of its processes. Mining and oil exploration are the heaviest and most destructive users, since they contaminate the water they use, which can then never return to source; but every area of industry exploits water, even where that is invisible. The silicon chip and the electronics industry it generated in California, for example, are among the heaviest industrial users of water. So we need to refine our image of water use to include the water that cools nuclear power stations, the water that drives electricity turbines, the water that gushes from oil drilling wells and so on.

The question is, where does that water come from? The available sources of fresh water are rivers, lakes and underground aquifers, which are replenished by rain through run-off. But that will happen only if it is allowed to happen, if the run–off is not contaminated, for example, or if the aquifers are given the time they need to recharge! A water crisis occurs when that process is disturbed, or when the hydrological cycle – the natural cycle of use and replenishment – is interrupted or blocked. But that is increasingly what is occurring as a result of human intervention in the water cycle, for example with pollution and its contribution to the process of global warming, or intensive agriculture with its overuse of groundwater.

Because water has been seen as an inexhaustible resource, wasting it on vanity projects has been a common way of displaying wealth. The water gardens which the Moors brought to Spain also served to display the power of rulers, but they were always functional as well as decorative, and treated water with respect and reverence, as they do for example in Granada's Alhambra. The same cannot be said for the giant column of water in the centre of Phoenix, Arizona (whose name is eloquent) or for Las Vegas, once the camp for workers building the Hoover Dam and now a city of 2 million in the desert. Dubai, a desert city crowded with luxury hotels and shopping malls, creates beaches and sea inside its buildings. They are representative of a massive migration over time from the banks of rivers or the coasts into desert regions. This only became possible as electric pumping was developed and it became possible to transport water. (California, for example, uses 15 billion kilowatt hours just to pump water).[15] As Jack Nicholson's character says in Roman Polanski's *Chinatown*, 'You can take LA to the water or you can take the water to LA.' There are, however, two major problems: the first is the impact on the water sources themselves and the second is the enormous amount of energy, fossil or otherwise, that is required to move the water in such quantities and over such distances. But its extravagant use is also because neither industry nor agriculture has considered water a cost, in the same way as labour, or energy or transport. As we enter the twenty-first century, it has become clear that its misuse will have enormous costs, both financial and social.

Wastewater and water wasted

Although it accounts for only 10 per cent of world consumption, most public attention has been focused on individual, domestic uses of water. Of the 1,700 litres we may individually consume, probably no more than 5 or 10 are for drinking; the rest is for the myriad domestic ends to which we use water – for washing, cooking, bathing and for removing waste, human waste. It is the case that most of that domestic water is in fact 'blue water', freshwater treated to a high standard for human consumption that is also used for other purposes – to flush toilets, wash the dishes, water the garden and so on. There are now waterless washing machines and dishwashers. In the USA and Europe low flow toilets are increasingly used, but there are many other ways in which wastewater can be recycled for human use, and even for drinking – as it is in Singapore. While the problem of polluted wastewater is critical, it has not been addressed in any major way – largely, probably, for reasons of cost. Yet there are now significant alternative ways of degrading human waste biologically, using water sparingly.

It is industry and agriculture, however, that use the remaining 90 per cent of water and who are responsible not only for the emissions that lead to global warming,[16] but also for the pollution and contamination that enters unrestrained into the water cycle and significantly reduces the water available to the population of the planet.[17] This is where the depletion of water occurs on a grand scale.

There are many ways of re-using water. The desalination of salt water is now being investigated worldwide, and used successfully in a few countries, though it in fact provides less then 1 per cent of available water. In oil-rich arid countries like Saudi Arabia, desalination plants provide much of the country's supply; nine of the world's ten largest desalination plants are in the Middle East. Thirty per cent of Israel's water comes from desalination, and a quarter of Barcelona's population gets its water from the Llobregat desalination plant. China too is developing desalination technology, as its water crisis intensifies. And it is increasingly seen as an alternative as freshwater has become saline because of the intrusion of sea water into rivers and waterways. But there is a major problem still unsolved; desalination uses enormous amounts of energy, as well as generating heavily saline brine which can have devastating effects on sea life. Wastewater recycling is an alternative; that has been most recently successfully used in Singapore, not only for non-consumption purposes but also for drinking. Other schemes, like rainwater harvesting in India and even the harvesting of rain in Peru and Chile, have generated new sources of water for agriculture.

But none of them can fully address the accelerating levels of water pollution by industry – chemical and pharmaceutical products leaked by mines and steel works and paper mills directly into rivers and waterways. Intensive industrial agriculture has boomed on the basis of the extensive use of pesticides and herbicides, which through run-off or seepage have also contaminated rivers and streams. In the developed world, and especially since the environmental impact began to be acknowledged, stricter environmental laws and controls have been introduced. The Clean Water Act of the USA, for example, requires the cleaning of coal before it is allowed to emit carbon dioxide (CO_2) and other contaminants into the air. The problem and the contradiction, however, is that this requires significant quantities of water as well as large-scale investment. For the developing world the growth of industry and export agriculture has brought with them all these problems, yet in the largest 'anchor' (or industrialising) economies like India, China and Brazil, what little environmental legislation exists is often ignored in the race to compete in the global economy, where their advantage can only be the result of cheap labour and the low cost of production – partly at least because there are so few costly environmental controls to consider. It is, however, human waste that is a key problem, the overwhelming bulk of which is simply discharged into urban sewers (at best), which flow directly into rivers and waterways, or into aquifers. The conclusion, then, is that the major problem humanity face is the *quality* of water, as well as its absolute quantities – although this too is an issue to be addressed.

Cityscapes: the absence of water

In the latter half of the twentieth century the cities of the developing world grew in a different way to their nineteenth-century equivalents, exploding through waves of often chaotic migration, expanding in unplanned townships and barrios around older city centres. Even where municipal organisation did exist, these new habitats could not be supplied in the same way.

The population of the planet is now 50 per cent urban. Of that number a rising proportion live in what are called the megacities, the huge cities of the developing world whose populations are counted in the millions and tens of millions. It is hard to imagine a city of 21 million (São Paulo) or 23 million (Mexico), let alone the 36 million that are crammed into the city of Tokyo. But the water crises (there are more than one) faced by these massive conurbations are not simply the result of size but of the manner of their growth. Most of them had populations at or around one million in the early 1950s. Their explosive growth has been a consequence of a series of interlinked factors. The dam building frenzy expelled perhaps 12–15 million people from their homes and villages; the growth of export agriculture has driven millions more from their subsistence plots to the mushrooming shanty towns on the hills surrounding the cities; the promises of modernity and of consumer culture drew many more towards the cities in search of work or paradise; hunger and thirst augmented by the drying out of rivers, or desertification, or the disappearance of fertile food plains and wetlands account for many more. The cities had no infrastructure to sustain or support these unpredictable columns of migrants, and government and municipal administrations found a hundred reasons to ignore their presence. The miracle is that these communities sustain themselves from their own ingenuity and the extraordinary creativity of those who struggle to survive.

The forests of wire attached to overhead cables, the jeeps that negotiate the steep earth slopes that climb the hills – the only vehicles that can negotiate them and provide a form of collective transport – the statistics for water theft that every city administration offers up as an alibi for its own inaction. All these are evidence of the ingenuity of the poor. For the most part, they do not appear in official employment statistics, either as employed or unemployed. Since they rarely qualify for state welfare – for which addresses and documents are usually required – they exist in a world that is euphemistically described as 'informal', or more appositely 'precarious', an economy whose scale is largely unknown but whose presence is as visible as the millions of street stands that sell everything from food and medicines to the imitation Rolexes and Nikes that are sold in this parallel but invisible economy that mimics the consumer culture of the burgeoning urban middle class.

Nothing so clearly marks the class divide in these societies than water. For most of the poor of the third world, water does not come in pipes but in irregular visits from tankers, or in small plastic bags from the back of lorries. In the 1960s, for example, rural migrants moved towards Mexico City and created a self-built city of ramshackle dwellings called Ciudad Nezahualcóyotl. It grew rapidly until its population passed the two million mark – but their water supplies came in lorries, and its cost could be dozens or hundreds of times the cost of municipally supplied tap water. In the West, anyone spending more than 3 per cent of their income on water is defined as living in water poverty; it is not uncommon in developing countries for the poor to pay 20–25 per cent of their income for water. It was a calculation of the cost of water under multinational control that sparked the Cochabamba explosion.

'Access' to clean water was the watchword of the Millennium Development Goals, with its promise to reduce by half the 1.2 billion people who had no potable water immediately available to them. Laudable as the aim was, it still left half a billion without access, in an era when huge plastic bags towed by tugs were transporting one and a half million litres of water across the seas – but only to those who could afford to pay for it. The town of Sitka in Alaska recently signed a 30-year contract to export 160 million gallons of its water per year to China. It was a potent testimony to the fact that water is now treated as a commodity, whose value is its price, and like all commodities, it will go to the highest bidder. What then of the 600 million still waiting for access?

A related Millennium Development Goal, less openly discussed because it deals with human waste – with shit in a word, and in vast quantities – is sanitation. Here the United Nations recognised 2.6 billion without access to proper waste disposal facilities. This rather anodyne expression veils a much more distressing reality – that the drains and gutters of the megacities are running with human waste as well as rubbish and the noxious products of modern living – plastics, industrial packaging, pharmaceuticals, the chemicals in discarded food and so on. Because the bulk of this waste is untreated – it is after all an expensive process to build waste treatment plants and they require considerable amounts of energy too – it flows directly into rivers and waterways, from where it enters the water cycle.

There it will join the poisonous pesticides and herbicides that are used in ever-increasing quantities in export agriculture as the first generation of these 'miracle' fertilisers lose their function and new generations of weeds and pests emerge with resistance to them. And they will merge in their turn with the perilous wastewater from mines and oil wells, not to mention the 'produced water' (how expert capitalism has become in procuring euphemisms), which is the radioactive product of fracking.

The right to hygiene is surely as central a human right as water. Yet water-borne diseases are responsible for 80 per cent of infant deaths in the developing world, according to the World Health Organization. Cholera, for example, was emblematic in the development of public health regimes in the nineteenth century. Municipal water provision was a direct result of the discovery of the relationship between cholera and the lack of clean water. By the later twentieth century it had become less and less frequent – until, that is, it made an unannounced reappearance in Peru in 1990 and again in KwaZulu, South Africa in the year 2000 and in Alexandria township a year later. Even if the UN's reduction of 50 per cent, as set out in the Millennium Development Goals, cuts the number of people without sanitation to 1.3 billion, it remains a twenty-first-century scandal. And while it is referred to as a 'crisis', as if were unexpected and unattributable, there are explicit and identifiable causes for this situation.

Some, it is true, have to do with the lack of infrastructure, but for the most part they are as Professor Asit Biswas of the Third World Centre for Water Management relentlessly insists,[18] essentially issues of water *management* as well as water quality. But there is a further key issue that in our view explains many of the water crises that have affected the planet and its people in the last 20 years or so.

It is not a coincidence that deteriorating water services have coincided with a critical ideological shift. From water conceived as a public good, or a *commons*, it began to be described in the early 1990s as an 'economic good', in other words as a commodity and thus by definition a good whose value was determined by price and which was available to be bought and sold. Strictly speaking, of course, a commodity is a product of human labour, material transformed by human intervention. Its price and certainly the profit that can be realized on it, is related to the amount of labour-time it 'contains'. There is a difficulty with the products of nature, like water – they are not *owned* or to use the more commonly used term today, they are not *private*. We discuss at some length below how the privatisation of water has brought multinational corporations into the water industry, and generated enormous profits for them. But the prior condition for that to happen was that water – and indeed nature itself – had to be *privatised* to become a commodity. Its availability had to be restricted somehow and its presence across the planet somehow enclosed, as the common lands were enclosed in the early stages of capitalism. In a sense, the privatisation of water is a second Enclosure, in which the earth is parcelled into property. Water, however, resists enclosure; it is anarchic and recognises neither fences nor national frontiers. So corporate capitalism redefines water as an economic good on the basis that it is treated and delivered to the consumer, at least in cities, and thus transformed into a commodity, the object of an exchange between owner and consumer. It is what bottled water has done on a massive scale – and with exploding profitability – even though it is tap water that is contained in the plastic bottle.

This process of privatisation that neo-liberalism has set in train is propelled by multinational corporations seeking new areas of investment. It is capitalism in its most recent and most predatory form. Though occasionally it may wear an environmentalist mask – by claiming that pricing water is a way of conserving it, for example – capitalism is blind to ecology. But it is worse than that. In transforming nature into commodities, it homogenises nature[19] and divides it into separate commodities. It not only degrades nature over time – polluting rivers, destroying wetlands, cutting down rain forests, diverting rivers – it also undermines the complex interconnected systems which sustain it. Plantations of trees in serried ranks of a single genus replace the mixed forests that regulate water flows, fertilise the soil and play host to an infinity of plants, creatures and insects. David Harvey calls it 'the wholesale commodification of nature in all its forms',[20] through a process of 'accumulation by dispossession' – 'releasing assets at very low or zero cost, providing immediate profitability.'[21] The application of these insights to the privatisation of water is immediately obvious.

HYDRAULIC SOCIETIES

Water has shaped societies throughout history; it has also determined their downfall. The great civilisation of Sumeria in southern Mesopotamia grew up some 7,000 years ago between the Tigris and the Euphrates. An agricultural society, it cultivated cereals, vegetables and pulses as well as raising animals, on the floodplains of the two rivers, using irrigation. Floods would often sweep away the channels or break them, and they would need to be frequently rebuilt by the ordinary Sumerians – though the wealthy were exempted. It was a system in which water was power, and those who controlled the canals and irrigation ditches ruled the social system as a whole. Where feudal Europe was based on a system of independent landed aristocrats, competing among themselves, the 'hydraulic societies', like Sumeria, were governed implacably by authoritarian rulers. Since water was the literal life-blood of the society, the structures that carried the water to the fields had to be kept in good repair, using the manual labour of the population. In the sixth century AD, the Abbasids built their huge irrigation projects in Baghdad and developed a diverse and sophisticated culture until its stagnation in the twelfth century as a result of failed water management. These were what the German historian Karl Wittfogel (1896–1988) called 'hydraulic societies' whose strength was the control of water and which fell, for the most part, when salination, excessive silting or floods and droughts undermined that control.

From the standpoint of the historical effect, it makes no difference whether the heads of a hydraulic government were originally peace chiefs, war leaders, priests, priest-chiefs, or hydraulic officials. Among the Chagga, the hydraulic corvée was called into action by the same horn that traditionally rallied the tribesmen for war. Among the Pueblo

▶

Indians the war chiefs (or priests), although subordinated to the cacique (the supreme chief), directed and supervised the communal activities. The early hydraulic city states of Mesopotamia seem to have been for the most part ruled by priest-kings. In China the legendary trail blazer of governmental water control, the Great Yii, is said to have risen from the rank of a supreme hydraulic functionary to that of king, [...] becoming the founder of the first hereditary dynasty, Hsia.

No matter whether traditionally non-hydraulic leaders initiated or seized the incipient hydraulic 'apparatus', or whether the masters of this apparatus became the motive force behind all important public functions, there can be no doubt that in all these cases the resulting regime was decisively shaped by the leadership and social control required by hydraulic agriculture. [...]

Among the Hill Suk of East Africa, 'every male must assist in making the ditches.' In almost all Pueblos 'irrigation or cleaning a spring is work for all.' Among the Chagga, the maintenance of a relatively elaborate irrigation system is assured by 'the participation of the entire people.' In Bali the peasants are obliged to render labor service for the hydraulic regional unit, the *subak*, to which they belong. The masters of the Sumerian temple economy expected every adult male within their jurisdiction 'to participate in the digging and cleaning of the canals.' Most inscriptions of Pharaonic Egypt take this work pattern for granted.

> Karl A. Wittfogel, *Oriental Despotism: A Comparative Study*
> *of Total Power* (New York: Yale University Press, 1957)

Wittfogel found the same patterns in Inca Peru and in Aztec Mexico, among others. They showed why the Chinese character for politics means control of water, and perhaps why Sharia Law is also centrally concerned with water. Wittfogel had been a communist until the Stalin–Hitler Pact of 1940 turned him into a fierce critic of Stalin's Russia. His purpose in writing *Oriental Despotism* was more than simple historical enquiry. It was to show how a particular power structure, centred on great engineering projects and dependent on the unquestioning labour of its people, oppressed and silenced by a bureaucratic leadership, would reproduce the authoritarianism of the original hydraulic societies. Stalin, as far as Wittfogel was concerned, was its contemporary embodiment.

2

How Water was Privatised

The message in the bottle

Nine billion bottles of water are sold annually across the world. It is an extraordinary figure, especially given that it was a trade that barely existed until the 1980s. Since then the brands have proliferated, competing with one another to find names that convey a sense of virgin nature, of clean uncontaminated sources far away from the urban clusters where the water is drunk. The early adverts for Vulcan, for example, showed a prehistoric man gathering water from a gushing stream. Other brands followed, emphasising that idea of spring water (Highland Spring, Poland Spring) and many more. The labels and the television and film advertising reinforced that idea of the natural, the uncontaminated, with landscapes of mountains and forest.

In fact, water began to be bottled in the mid-nineteenth century in the United States.[1] Poland Water first appeared in 1845, followed by Vittel (1855), Deer Park (1873) and Arrowhead (1894). The poor reputation of the water supplies in New York and other cities, plus the growth of railways, spurred their creation. Nonetheless, the quantity of bottled water sold was still negligible even up to the 1960s. Then came Perrier, first manufactured in 1976.

Perrier's achievement was to turn water into a social drink, consumed in elegant bars by elegant people watching the bubbles coat the ice cubes and the lemon; it became a kind of fashion code. It is hard to pinpoint the critical selling point, but the timing tells us a great deal. It was in the late 1970s that arguments about pollution begin to occupy the media and to preoccupy social commentators. Environment entered the vocabulary and with it a succession of images of a natural world very different from the expanding cities of both developed and developing world. The water bottle bridged the two and promised health, a source close to nature and a drink free of the harmful effects of alcohol or coffee. The plastic bottle also coincided with a health boom, the widespread addiction to running, jogging and marathons, each of them associated with the throwaway water bottle. The images accumulated as a kind of reassurance of the availability of health without sacrificing modish consumption. And it worked – hence the billions of bottles of water sold across the world, in areas where there were public supplies available to almost everyone.

The irony, of course, is that not only was this healthy, natural water no safer than tap water – which is subject to regulation in a public health context. In possibly up to a quarter of cases, it *was* tap water that was contained in the famous plastic bottles, as the case of Coca-Cola's brand 'Dasani' revealed – it had been taken directly from the public water supply![2] Tests carried out by the FDA in the United States, and in Holland, furthermore, found that something like 40 per cent of water bottles contained contaminants of one kind or another. And to pile up the ironies, the bottles themselves may have carcinogenic properties,[3] as well as requiring up to 100 litres of water for their manufacture! But the most significant issue of all is cost. Bottled water is far more expensive than tap water – in fact, it is far more expensive than petrol at around $8 a gallon.[4] Barlow and Clarke estimate that the cost of bottled water can be 10,000 times the cost of the equivalent tap water!

In many countries in the developing world, there is no reliable public water supply, and most people obtain their water from water sellers or tankers at an equally inflated price. In the major cities of the developing world, the patterns of growth and expansion were often chaotic and unplanned as well as unregulated; there were neither the resources nor the opportunity to lay pipes to supply water beyond the original city centres. What water provision existed, was therefore under severe strain and often reliant on a very old infrastructure. The leakage figures drive the point home dramatically. One result of this has been that the sales of bottled water have expanded into the cities of the developing world, though at a cost which ensures that they will be accessible only to the wealthier classes, who are by and large deeply sceptical of the safety of public water supplies. For the poor, water will arrive with absolutely no health guarantees with the irregular visits of water tankers, whose tariffs are far higher than the inadequate municipal suppliers. A further irony is that the cost of producing the plastic bottles and of transporting them contributes significantly to global warming, a further cause of water crises.

The so-called Perrier Revolution was, to say the very least, ill-named. It was not a revolution, but a massively successful commercial campaign to create a new mode of consumption, and a new commodity, which generated in its turn enormous profits for very little outlay. It was a skilful play on public fears at an early stage in the transformation of public culture that we now know as neo-liberalism. For it had at its base a suspicion of public provision consciously and deliberately created to spread fears and doubts in anticipation of wider privatisation. This went hand in hand with an ideology of independence and self-reliance, of radical individualism, which encouraged private consumption and reinforced a none too subtle attack on what came to be called a 'dependency culture', as if relying on public services made parasites of us all. The first example of the privatisation of public water supplies occurred in the UK at the hands of Margaret Thatcher, closely followed by Chile.

The manufacturers of bottled water came largely from the food and drink sector, where the giants of the industry – Nestlé, Coca-Cola, Pepsi, Danone – prevailed. The effect was to bring increasing pressure to bear on municipal water authorities, using the argument that they were using public funds to maintain a monopoly. It is, of course, a fraudulent argument – public control of the commons is a guarantee of democratic provision, though it is not by any means a guarantee that those who run the industry will be exempt from inefficiencies or mismanagement. But that is an argument for more public oversight, not for privatisation, the result of which, as we shall see, is the creation of non-accountable international monopolies which have exploited public enterprises for private gain. And at a huge profit!

The privatisation trap

The full transfer of public water provision to the private sector began in Britain in the early 1980s under Margaret Thatcher's direction. Like all processes of privatising the state sector, it was a bonanza for private enterprise. The Water Act 1989 privatised the ten regional water authorities in England and Wales (Scotland and Northern Ireland were not included in the process). All the regional authority debts were written off, and the purchasing companies were paid £1.6 billion as a start-up fund. Contracts were for 25 years and there was no competitive tendering, nor were the private companies regulated by the public sector water commissioner Ofwat. It would be hard to imagine a more favourable environment for private capital. And the results – price increases twice as high as inflation, and total profits over ten years of just under £10 billion – clearly caused some celebration in the boardrooms of the acquiring companies. Similar privatisations elsewhere in these crucial years produced a very similar profits boom. In 1990, for example, the Peronist government of Argentina effectively privatised the whole public sector, selling state enterprises at bargain prices to multinational companies from across the world.[5] The profits from this monster bargain sale (some $9 billion) were not returned to the Argentine state, but pocketed by some of the auctioneers!

The Dublin International Conference on Water and the Environment in 1992 proved to be a critical crossroads, encouraging a participatory approach to the management of this 'finite and vulnerable resource', but at the same time redefining it as an 'economic good'.[6] At the time, this may have passed without comment – but in retrospect, it is clear that this was an ideological milestone, and a legitimation of the process that would in a single decade transform the provision of water. In a globalised age, privatisation also pointed to the internationalisation of the control of the commons, dealing a further, possibly fatal blow to local provision and any possibility of democratic control of this most fundamental of resources. Or so it seemed, at least, to the enthusiasts for the free market, like Thatcher.

The water industry until the 1980s was occupied by a number of local and regional companies, gradually merging across the planet into larger and more powerful entities. But the water business was dominated by two French companies, who by the end of the century controlled two-thirds of the world's privatised water market.

Vivendi, now Veolia Environment, employs 318,376 employees in 48 countries. Its revenues in 2012 were 29.4 billion euros. It has a dominant presence in the US water industry, covers 50 per cent of France's water provision and supplies water to 110 million people in 100 countries. And it is about to enter the Chinese market. Its competitor, though the two corporations often collaborate in joint ventures, is Suez, formerly Lyonnaise des Eaux. A third, but significantly smaller member of this very exclusive club was German, RWE, though it has since been sold to an Australian based multinational. (Biwater, Saur, Anglian Water and others follow behind). Bechtel (International Water), a massive engineering company which built the Hoover Dam, would reappear in another guise in Cochabamba, Bolivia, where its attempt to take over the regional water provision was reversed by mass action.

The massive expansion of the private water industry was in a sense the distillation of neo-liberalism's commitment to the privatisation of all public services and public goods. The two companies had begun their drive to expansion and corporate growth in the 1980s in France and later Italy, winning or buying allies in the municipal water sector as it was prepared for transfer to the private sector. In 1985 Vivendi (as it then was) signed a lucrative contract with the office of the Mayor of Paris. Between 1989 and 1995, it won 3.3 billion euros worth of contracts, part of which was then diverted to political contributions. In 1996, the director general of Vivendi was sentenced to 18 months in jail and fined 27,000 euros for illegal commissions to state employees like the mayor of Angouleme. Other similar accusations followed. Not to be outdone, it was Suez that was the focus of the most high profile of these cases, when the mayor of Grenoble was jailed for accepting large bribes. This is more than anecdotal; the Argentine case, among many others, like the cases involving the French corporations, illustrates that they were part of a global phenomenon – the massive transfer of public funds into the private sector, another example perhaps of David Harvey's concept of accumulation by dispossession.

What is very clear is that the transfer was effected at extraordinary speed and with little resistance in its first phase during the 1990s. Vivendi's water revenues leapt from $5 billion in 1990 to $12 billion in 2002 and, as we have seen, continued to rise exponentially as the decade progressed. But this was not, as the company websites might have us believe, thanks to the ingenuity and insight of the company's directors.

> The loans from international development institutions, which are all public sector, government institutions, are central to the financing of nearly all the privatizations. Private investment from the multinationals themselves may be only a small part of the money involved.[7]

Throughout the 1990s and thereafter, as globalisation progressed, individual states and governments were put under pressure to privatise their services and effectively dismantle the state, the better to facilitate the free flow of multinational capital. Cash-strapped municipalities and governments were offered loans from the World Bank, the International Monetary Fund, the international and regional development banks, to which were attached 'conditionalities'. We know by now that such neologisms invariably hide something sinister – and this was no exception. Effectively the loans were conditional on privatisation, sometimes veiled as a Public–Private Partnership (PPP), which effectively entailed the privatisation of the public sector or at the very least its subordination to private sector priorities.[8] But it became very clear very quickly that the public partner would be expected to maintain much of the infrastructure, while the private companies enjoyed the profits from water distribution and supply.

Figure 2.1   Testing the waters of the Yangtze for contamination.
© Lu Guang (Greenpeace)

The assumptions on which 'conditionality' was based were ideological rather than in any way objective. It was true, for example, that the infrastructure of the water industry was failing in many places. In Britain, privatisation was preceded by dangerous contamination of water supplies in the south-west of the country and the failure of supply in some authorities in the north. Figures for a number of major cities show an enormously high level of leakage and infrastructural failure. But the implication that the multinationals would address problems of infrastructure was ill- founded. The defects of municipal utilities had to do with the progressive reduction of public funding as much as with management failures. But the privatised companies were concerned only with the delivery of water and the profits to be derived from that service; the conditions imposed by the World Bank and others, and reinforced by WTO regulations, made no demands of the new owners in that respect. So while Veolia, Suez and the major water suppliers increased the number of homes they jointly supplied from 51 million in 1990 to 331 million in 2005, their profits reached around $40 billion annually in the same period. Yet the problems of infrastructure remained largely unresolved. In Buenos Aires, for example, Suez failed to fulfil its maintenance commitments yet realized profits of 19 per cent after tax between 1994 and 2000, before pulling out three years later because the business was insufficiently profitable!

The argument advanced in favour of privatisation was that it would drive forward the solution of problems and on simple market grounds bring water where it is needed.[9]

The test of privatisation and its claims to efficiency was whether it could deliver water to sections of the population without access (essentially the poor), maintaining or raising water quality and keeping prices at an acceptable level. The implicit assumption, of course, was that private capital would invest (albeit for profit) where the public sector had failed to do so as it was starved of resources. This was the utopian version of neo-liberalism – that the drive for profit would impel investment and improve the living conditions of all as a result.

Even before the bonanza of the 1990s, the main players in the water field were associated in the public mind with the corruption of public officials. Their budgets were huge and the potential pickings were rich – and in any event, the public sector was being deliberately neglected. Furthermore, public sector water managers would often slide effortlessly into equivalent posts in the private sector at far higher salaries.[10] At the same time there is a constant movement of personnel between the executive committees of the multinationals, governments and the members of the various commissions charged with reviewing water provision on behalf of the World Bank and the IFC (its financing arm for water projects) as well as the UN.

There is very clearly a revolving door between the private companies and the international institutions and governments.[11] A critical analysis of the impact of privatisation is unlikely to come from there or indeed from any of the international organisations which they run and control. In

2004, Kofi Annan, then Secretary General of the United Nations, set up the UN Advisory Board on Water and Sanitation led by Michael Camdessus, previously of the IMF, and including several executive members from Suez. What is clear is that where there is an urgent need to address water poverty – in the poorer countries of the developing world for example – the private monopolies have shown very little interest. The aftermath of privatisation is invariably an increase in the price of water. In Argentina, for example, where the privatisation of Aguas Argentinas was emblematic of the new neo-liberal order, prices were raised in 1991 *prior to* privatisation in order to attract bids. Yet ten years later, in 2001, the Suez-led consortium that had taken over the enterprise was fined for overcharging. In Tucumán, Argentina, where water was privatised in 1995 (under Vivendi's subsidiary, Aguas del Aconquija) with a 30-year concession, tariffs were doubled yet no investment was forthcoming when the water turned brown. The public refused to pay their bills and the concession was ended in 1998. In Chile, the public company EMOS – which had been highly praised by the World Bank for its efficiency and coverage, was privatised in the face of public resistance. In the following year rates rose by 15 per cent, producing an increase in profits for the new company of 197.6 per cent.[12]

The IFC boasts that it funded 847 water projects between 1993 and 2013, half of them in Latin America. Their record in terms of public service is disastrous. But then they have no public service commitment. The head of Vivendi was brutally clear: they would not provide services to the poor without government subsidies and investment from the public sector.[13] The myth and the reality move then in opposite directions. The claim that the private sector would support public provision proves to be the reverse of the truth; public funding must support the private sector, as if the financial support of publicly funded institutions like the World Bank were not already enough! Suez, for once departing from its shared script, claimed that it would extend water supplies to the poor sectors in the cities. It did nothing of the kind. In Manila where a 25-year contract worth $2.7 billion was awarded in 1997, the record was equally appalling. Leakage hovered around 50 per cent, while prices have been repeatedly raised. The IFC's insistence that privatised provision costs 20 times less then the prices charged by the water vendors pales when we recall that their prices were several hundred times higher than tap water to begin with! In Lagos the relentless pressure to privatise has produced several plans, each rejected by the municipal government. The

1999 IFC proposal was dismissed by them as 'ridiculous'.[14] No agreement has yet been reached, and the majority of Lagos's poor still buy from the street water sellers. And finally, there is Jakarta, Indonesia, the largest privatisation contract ever awarded. Amid repeated allegations of corruption, water tariffs increased fourfold after 1997. The privatised company's prices were 2.7 times those charged by the public utility in the second city of Surabaya. The contract was terminated in March 2015.[15] The water corporations have largely pulled out of areas such as the South African townships, on the grounds that the poor cannot pay. The result, despite the inclusion in the South African constitution of the right to water, is that fewer people have access to clean water in 2004 than ten years earlier.[16] The laws of the market are nasty, brutish and very clear.

The rush to privatisation under globalisation has yielded enormous profits for the dominant corporations, supported and financed directly or indirectly by international financial institutions on the basis of a false claim that services would improve and reach the needy. The evidence that it was false is overwhelming. According to Anne Lappé of the Small Planet Institute,[17] 34 per cent of water and sewerage privatisations have failed, compared with just 6 per cent in the energy sector and 3 per cent in telecommunications. Most significantly, however, by the second decade of the twenty-first century 180 cities in 35 countries had re-municipalised their water and most had turned to interest-free Public–Public Partnerships (PUPS), sharing advice and operational experience, as well as finance with other municipal utilities.[18]

According to the IFC, 768 million people still have no access to drinking water, 2.5 billion are without sanitation and nearly 3.5 million die every year from water-related diseases.[19] Mexico City provides the most devastating example. The city was built on a lake drained by the Spanish conquerors where the Aztec capital of Tenochtitlan, a city built on water, once stood. Today one-third of the Mexico City's water leaks into the ground at a rate of 12,000 litres per second. The city water used to come from the river Lerma, now exhausted, and now comes from the once beautiful Lake Chapala in the state of Jalisco and from the river Cutzamala, depriving the local indigenous nation, the Mazahua, of their water supplies. Meanwhile an extraordinary 95 per cent of the city's population suffers from gastro-intestinal disorders, the result of the irrigation of the main fruit and vegetable growing area in the neighbouring state of Hidalgo with raw, untreated sewage.[20] The city's water was privatised in 1980.

What has been the pattern of water privatisation? The astonishing part of the story is the speed of growth of these huge multinational enterprises whose turnover annually exceeds $40 billion – and the projection for future decades is that it will rise to $100 billion. Yet these huge profits from water have not contributed anything to the resolution of the water problem; the beneficiaries have been the company executives who move effortlessly between water advisory bodies, the World Bank and the IMF, and the boardrooms of the major companies. The paradox is that only some 5 per cent of water provision in the world is in their hands directly; in most of the world where municipal water providers exist, they remain in place. The multinationals – the Vivendis and Suez of the world – derive the lion's share of their profit from service contracts and distribution deals. Their enthusiasm for assuming direct control of the water itself waned dramatically when the real state of the infrastructure became clear. The pipes and pumps that distributed the water tended to be old and inefficient. The figures for leakage tell their own story. The cost of water to the domestic consumer is probably almost doubled by the waste and loss represented by leakage. For profit-hungry companies, of course, repairing the pipe network is of no interest – there is little to be gained from that. On the other hand, transporting water – by whatever means – generates considerable earnings. And it has a further advantage – that it is only the rich countries, who can afford to buy their water from the other side of the world, in order to use their own water for more lucrative investments. China is now one of those who can import potable water, in this case from Alaska on a 30-year contract. More importantly, the very presence of these giant water corporations is evidence of the pressure from industry and agriculture in particular that has raised the demand for water to the point where the possibility of a crisis in water provision has become an issue of public debate. Thatcher's privatisation of water revealed how much there was to be made from the strategy. All the multinationals moved into Britain's water industry very quickly – there was very little risk given the direct financial support, exemption from taxation and sweetheart deals they were offered. It was enough to excite the water companies worldwide. China, with its permanent water shortage, has now employed Veolia to provide water services for several major cities. It is simply the latest in a series of shared projects with multinational corporations. But it presents a double irony; first, that it should be calling on ruthless capitalist corporations in the search for a solution to its water problems. But beyond that it is the unregulated acceleration of industrial growth and the uncontrolled use of pesticides, which China now exports, that has made so much of China's water undrinkable.

The principal argument for privatisation was that the market would provide more efficiently, more cheaply, and reach the remotest parts of the planet in search of markets. The reality has proved to be very different. First, the presumed inefficiency of the public sector was simply untrue; no evidence was ever provided for the argument, because in a sense it was a well-disseminated myth. There was inefficiency, mismanagement, corruption and laziness in the public sector just as there was in the private sector; the difference, as we have discovered to our cost, is that the public sector was always accountable in the last resort to society as a whole – the private sector had no such trammels or limitations, as the financial crisis of 2008 and subsequent revelations (Swiss Leaks being only the latest) have so dramatically exposed. The Chilean example may serve, since its public water enterprise was efficient and effective, yet it was privatised – with the usual effects of rising prices for water and increasing problems in its delivery.[21] The two decades of privatisation have produced no significant improvement in services anywhere, and the cutbacks in state funding for water, compensated later by a range of special arrangements and financial sweeteners for the multinationals, have produced only a bonanza for shareholders and enormous bonuses for directors, all derived from higher prices to consumers across the board. And for the populations most in need of water, from Palestine to South Africa, who it was argued would benefit most from the Millennium Development Goals, the cost of water has risen, at the expense of other necessities. That is the reality of privatisation of water services.

Privatisation represented a fundamental ideological shift in favour of multinational capital, undermining the philosophy that had governed the provision of water from the mid-nineteenth century onwards in the countries of the developed world. The arguments offered by Thatcher and those who followed her – for example in Chile in 1999 – was that municipal provision of services, including water but also housing, education, health – was inefficient, badly run and excessively costly. Removing the state from the provision of public services would allow the market to offer them as and where they were needed. The British Water Act of 1989 coincided with the publication of Francis Fukuyama's *The End of History*,[22] which generalised the argument in the wake of the fall of the Berlin Wall. Even Fukuyama, some 15 years later, was forced to acknowledge that the promise of a global market providing for the needs of all was an ideological fantasy. Yet by then, the pursuit of profit as the sole guiding principle of economic life had already wreaked its destruction – in Iraq, in Afghanistan, across the developing world where public services were wilfully dismantled and destroyed and delivered to a private sector which recognised neither social need, public accountability nor the right to the basic conditions of life. Nothing symbolises the significance of that shift more clearly than the issue of water provision. It would take a decade before the struggle to reverse the process and restore public services and public spending would begin in Latin America and across the world. In that time, the water multinationals exploded in size and diversified their activities; the profits they made from water, meanwhile, would be invested for profit in the production of other commodities, and not in the provision of water itself. Perhaps there was some awareness that people would soon begin to see through the fraudulent promises of privatisation, and its accompanying argument that water was a commodity like any other, and demand that it be reclaimed as public good.

As to the providers of bottled water, their profits have rocketed too. The extraordinarily high cost of bottled water compared to tap water has created a huge consumer market, not only in the West, but also in Africa and Latin America, where it is the wealthy few who benefit from piped water and the poor who must have recourse to the bottled variety. The much trumpeted health benefits from bottled water are, of course, both a lie and a diversion. A lie because tests in the USA and Holland have found up to 40 per cent of bottles tested were contaminated with E-coli and other bacteria;[23] bottled water is not subject to the more rigorous standards applied to public water supplies. And a diversion, because the issue in the developing world is not simply the provision of drinking water but the adequate provision of water for sanitation and sewage. The failure of privatisation has, as is the way in capitalist economies, brought new opportunities to other capitalists – in Mexico City where, as we have seen, the quality of the water is declining even from its very poor beginnings, only bottled water (and only some of that) is reliable. In Delhi, according to Asit K. Biswas, nobody trusts the water, and bottled water sales are rising. The rushing mountain streams on their labels will soon be flowing no longer.

There is a striking image to represent the moment. In Iraq, after the US invasion, when water supplies like all public services were fatally disrupted, the local populations queued for water at street stand pipes. The US troops watching over them were supplied with ice-cold bottles of Poland Water in unlimited quantities.

Beyond that, the large-scale pollution of water sources – rivers, lakes and aquifers – is a more central problem still unrecognised in international debate. California, for example, has recently announced that it will close down 120 oil wells,[24] which have had permission to discard their waste products directly into a water basin. If they are in fact closed down it will be a small step forward, leaving only a further 2,280 wells to control! And that is in the state of California alone.

The privatisation of water distribution is not, however, the whole problem – it is only part of it. What the US example shows with startling clarity is the fact that the owners of water are the owners of land. The aquifers, for instance, lie beneath land that is by and large privately owned; and the laws of capture, which govern most water ownership issues outside the major cities, in fact give the landowner the right to sell water. To state the very obvious, it is of course impossible to parcel off water in underground basins. So capture actually involves taking and selling water from the *whole* body of water rather than from your specific area. The startling depletion of the massive Ogallala Acquifer in the USA illustrates the problem very clearly. When Los Angeles bought the Owens Valley it bought the water beneath it; the result is that the land in the valley is contaminated and risks, with a tragic irony, becoming itself a dustbowl.

The paradox, if that is what it is, is illustrated in hundred of other examples from across the world. In India, for example, the Coca-Cola bottling plant in Kerala became a focus of protest and resistance from local farmers as well as symbolic of a wider misuse of water across India by large multinational concerns. The rising level of suicides among poor farmers is a desperate response to the colonisation of their water by enterprises like Coca-Cola, the limitation on farm subsidies to small farmers while the giant sugar and cotton companies receive generous state support. Faced with the diversion of their water, those small farmers who remain have dug deeper and deeper wells, directly into the aquifers. In Latin America, the purchase of tracts of land by mining and oil companies has given them control of water resources which in many cases are returned to local rivers and lakes contaminated by chemicals and metals, while the intensifying use of pesticides also pollutes rivers, lakes and aquifers.

Privatisation, therefore, goes beyond ownership of water itself to control of water for other areas of production. Inevitably, it will take the issue into matters of global warming and into the complex issue of 'virtual water' (which will occupy us in Chapter 3). But having addressed the issue of where the water is and who controls it, we have then to turn to the wider framework. It is not enough to insist on the necessity of controlling water, nor even to tell dramatic stories of an end of the world scenario. Global warming is not distinct from the question of water usage; the Amazon experience is perhaps the most dramatic and at the same time the clearest evidence of that. The replenishment of the Amazon Basin, the maintenance of river flows and wetland integrity, are threatened by deforestation, which is growing at an exponential rate, but also by the purposes to which the cleared land is dedicated. The clearing itself emits CO_2 which would otherwise be absorbed by the forest and methane from the decaying vegetation, both significant contributors to global warming; but the massive soya plantations and cattle pastures that replace the disappearing cover also contribute directly to both global warming and water depletion. The former by the use of pesticides and chemicals, the latter by the amount of water used in both agricultural and industrial production. Anthony Giddens expresses his concern that:

> the natural sinks of the earth are losing their capacity to absorb greenhouse gases. Most climate change models (he argues) assume that some half of future emission will be soaked up by forests and oceans, but this assumption may be too optimistic.[25]

We share his concern.

3

Disasters, Natural and Otherwise

Floods and famine, storms and cyclones – natural disasters all, the vengeance of nature. That, at least, is how we have learned to see these events. More recently, and often reluctantly, it has become clear that many of these 'natural' events are the result of human intervention, human error or human arrogance. In the effort to contain or harness nature, or to use it for other ends, a delicate balance can be broken – and other kinds of tragedy follow.

Water can be life-giving, reassuring, comforting; it can also be fearsome, threatening, volatile. But it is probably at its most dangerous when it is transformed into a means of controlling others or as an instrument of power and ambition.

The thirst of Palestine

Not all water scarce countries are water poor. The clearest examples are California (not a country of course but bigger than most) and Israel. Both offer clear proof of the adage that 'water runs uphill towards money'.

Israel is a desert country; it depends for its very survival on water, and its relations with its neighbours have as much to do with water as with geopolitics. As Chaim Weitzman wrote to Lloyd George in 1919, 'the whole economic future of Palestine depends on water for irrigation, and electric power'.[1] In real terms, then, Israel's very existence will be at the expense of others. Its own water resources, calculated at 500 cubic metres per person in 1995, place both Israel and Palestine within a category of water scarcity. Sixty per cent of its water prior to the 1967 Six-Day War came from the Jordan River Basin, with 75 per cent belonging to Jordan. Its occupation of the Golan Heights on the border with Syria in the course of that war, an occupation which continues to the present day, gave Israel control of the headwaters of the Jordan. Two months before the war began it had bombed the canals built by Syria with the intention of diverting Jordan waters; it later blocked dams built by Jordan and it is almost certain that the occupation of the West Bank was to prevent drilling of the West Bank Aquifer, whose consequence could have been the salination of water supplies on the other side of the river. Today, under Israeli siege, the million plus inhabitants of the Gaza strip average 44 litres of water per day per person – the World Health Organization minimum is 100 litres.[2] In fact the Coastal Aquifer supplying Gaza is so contaminated by waste from Israeli settlements that 95 per cent of its water is unfit for human consumption. This leaves the Palestinian population with a single source of water – the water tankers, which charge 400 times the cost of directly piped water, and whose supplies are heavily contaminated. While the overall Palestinian average is 90 litres, in the areas under Israeli military occupation like South Hebron this may be no more than 20 litres per day. The Palestinians, who comprise 19 per cent of the population of Palestine, had access to only 2 per cent of the water resources. The Israeli Water Commission controls the water supply, which is why some 98 per cent of the river Jordan's water today is under their control.[3] Israel's high technology agriculture and rapid industrial growth were inconceivable without water, with the kibbutzim claiming over 40 per cent of the total for its intensive agricultural projects. This must represent a water war, and a particularly brutal and inhuman one, especially bearing in mind that these appalling figures we quote are from 2012, before the latest round of attacks on the Palestinian territories. Let this stand as an extreme example of water as a weapon of domination. But perhaps the gravest part of it all is that, in its commitment to making the desert flower, Israel has developed highly advanced techniques of drip-feed irrigation[4] and desalination which could benefit the entire region, as well as some original methods of transporting water – in their case, imported from Turkey in one and a half million one-litre plastic bags towed by tugs. But for the moment they remain weapons of war, and the Palestinian people are held in thrall by the cynical use of thirst.

Signs of things to come

It sometimes seems as if for millennia the water picture, having rolled through history in slow motion, has suddenly gone into overdrive. And it is not simply because the particular problems of water, and the general questions about the environment have suddenly emerged as the result of a collective crisis of conscience. There are certainly critical markers in the course of public awareness; Rachel Carson's *Silent Spring*,[5] a dystopian vision of a world that has for too long allowed the environment to deteriorate and slowly die; the Bhopal disaster in 1984 when a spill of toxic chemicals from a Union Carbide factory in India caused and continues to cause the death and sickness of tens of thousands of people; the tsunami of 2004; the massive leak from BP's oil platform off the coast of Florida among a growing list of man-made catastrophes. The awareness of the context in which we live has certainly grown, and has influenced an expanding public debate. But beyond that are the dramatic objective facts – that the water which every generation until now had largely taken for granted is now under threat on several levels; the supply of water itself is no longer guaranteed, as it was once assumed to be. And that is not simply because there is less clean water available, but also because the allocation of what there is today is determined by the market rather than by a concept of the rights of the citizen and the duty of the state. Furthermore, the quality of available water is increasingly declining as contamination and pollution are discovered in growing streams of water. While drinking water is still not directly accessible for over a billion people, and sanitation is inadequate for double that number, the disturbing reality is that water provision, or the lack of it, is increasingly determined both individually and collectively, by income levels. Whatever else we lacked, it was surely part of our very existence as human beings that air and water were rights not privileges – unless life itself is a privilege, and not a right. And it is a brutal irony that the projected scarcity of water (for as of 2015 there is still enough for all on the planet, or there would be if it were fairly and equally distributed) is seen by the powerful forces that dominate the global market as a commercial opportunity, in that those who can afford it will pay rising prices for this newly defined commodity, while those who cannot, will continue to live on the knife edge of water scarcity.

We have avoided catastrophic visions and predictions thus far because it is our concern to provide information and inspiration to all those who are fighting the commodification of water and for its recognition and treatment as a human right, a common shared resource. Gloomy prognostications tend to generate a sense of impotence, the feeling that this future is written as destiny 'eat, drink and be merry, for tomorrow we die'; or they are ignored.[6] But since food and water are interwoven, the pleasure they bring may yet have to be fought for by the majority of mankind. The difference here, perhaps, is that – as we will go on to argue, with examples – there are ways to act in the present to directly transform that future.

We have argued that the situation in which water is becoming seriously at risk has accelerated dramatically in recent years, and made it a much more urgent issue to reorganise the way it is used and distributed. There is still sufficient water to address the needs of the world's current population – just. Looking ahead, however, to a not too distant future – say 2025 or 2030 – against the background of population trends, it becomes clear that were things to continue as they are, we would be facing a very real crisis of provision. What has held human society together is the water cycle – 'the combination of natural physical, chemical and biological processes that constantly recycle water, ensuring a steady supply to support life on Earth.'[7] According to some predictions, by 2025 two billion people could be facing the awesome prospect of water scarcity, that is 'the lack of sufficient water to maintain current per capita levels of food production and meet expanding urban demands for water'[8] – and this bearing in mind the vast differences in living standards across the planet. What in California may signify the sacrifice of a private swimming pool, in much of Africa may mean the difference between life and death. There are some for whom that moment is already here. In Yemen, for example, whose capital Sana'a is predicted to be the first capital city to run completely out of water, and where the town of Taiz is restricted to one day of water a month. We have already seen that the one million citizens of Gaza are limited to between 20 and 50 litres per person per day, when the World Health Organization argues that the absolute minimum consumption to sustain life is 25 litres and the normal requirement is 100 litres daily to cover hygiene and sanitation. In both North and Sub-Saharan Africa available water in many places barely reaches that minimum either.

It would be easy, and convenient, to ascribe this situation to the vagaries of climate and geography. Yet for centuries even desert peoples have found ways to maintain their necessary supply of water. Basil Davidson, for example, in the opening chapter of his book on Africa, looks back at a green self-sustaining community in Nubia in what today is the Nubian desert;[9] and Constance Hunt quotes Fred Pearce's 1992 visit to the Nabateans, who began to farm the Negev desert about 2,000 years ago.

Pearce visited one of these ancient farms one spring after six weeks with no rain and found that the soil was damp, a field of wheat was growing fast, almond trees were in leaf and pistachio trees were budding.[10]

Clearly, then, it is human *intervention* in the environment in recent times that has generated the present crisis. The wealthier nations of the world, generally speaking in Europe and North America, harnessed nature to material advancement, industrialisation and modernisation, with their accompanying urban growth. And it was not only nature that they exploited, but labour, and this on a global scale beginning with the first colonisations in Latin America. Spain's wealth was clawed out of the mountain mines of Bolivia, Peru and Mexico, from the Amazon forest from earliest times, from the henequen fields of the Yucatan, from the 'devil's excrement' which burst from the ground near Lake Maracaibo in Venezuela,[11] from the sugar plantations of the Caribbean and the banana plantations of Central America. Britain built its empire on cotton, jute, and the cocoa plantations of West Africa. And it was the labour of enslaved and oppressed peoples that drew the wealth from the ground.

What has this to do with water? It is the point of connection between historical and contemporary exploitation, for water provided the energy that drove these systems of primitive production – and then as now what was extracted was taken from the impoverished, 'underdeveloped' world to enrich the developed, imperial centres of industry. And then as now both materials and human beings were taken to those centres, as poor immigrants, to produce the wealth of which they would never be the beneficiaries. In the early twenty-first century, water is taken from the poorer parts of the world to the wealthy centres of the global market both in the form of 'virtual water'[12] or, increasingly, directly in tankers or giant plastic bags.

The hidden cost of dams

If water was essential to driving the mechanical engines of the industrial revolution, just as the water wheel impelled the progress of pre-capitalist industry and agriculture, the amounts required grew exponentially in the modern industrial age. Cities grew, their populations expanded, and their food needs grew correspondingly. Yet the agricultural land that supplied them could, if drought or soil erosion took hold, become dustbowls, producing nothing but hungry immigrant populations. It was a spectre that haunted the North America of the late 1920s,[13] and one that lay behind the great engineering project of the early 1930s, the Tennessee Valley Authority and the Hoover Dam. It was not the first great dam in the United States, but it was the largest of its day and the most iconic, the expression of a New Deal that would mark the end of the Great Depression. It was also in many ways the beginning of modern water engineering. The contractor was the United States Bureau of Reclamation, which became, with the dam's completion in 1936, the largest water wholesaler in the USA, and remains so. But the six engineering companies who built it included one, Bechtel, whose name would become particularly familiar to Bolivians in 2000.

Despite the opinions of John Wesley Powell,[14] the dam was built; its main beneficiaries were the agricultural industries along the Colorado, and the California farming lobby. In the official allocation of the hydroelectric energy released by the dam, California was the largest recipient, receiving just over 25 per cent of the total. Thus began the era of dam building – a total of 78,000 of varying size were built across the USA by the 1970s. The Hoover Dam transformed US agriculture, but by placing enormous power in the hands of large farming interests, which was certainly not Powell's intention. This was a very different concept from *riparian rights*, which accorded shared access to water to all those living on the banks of rivers, or from the ancient *acequia* system which operated in the American south-west under Spanish colonisation.[15] Devon G. Peña, an acknowledged expert in the field and president of the Acequia Institute, describes it thus:

> Under customary acequia law, the irrigator, or *parciante*, does not 'own' the water as private property but rather has a right to use adjudicated water rights as an asset-in-place and only under a farmer self-managed allotment scheme based on the equitable and fair distribution average of the net cubic feet per second (cfs) flows in the ditch proportionate to the size of the private land s/he irrigates.[16]

Current studies of the acequia system confirm its efficiency, its conservation of the soil, as well as its egalitarianism. The Colorado pacts, by contrast established the principle of 'water capture' or 'water by appropriation', defining water rights as deriving directly from the ownership of land and that prior ownership determined absolute water rights. This became the dominant principle in most of the USA.[17] It meant, of course, that landowners could sell their water on to other interests, even where those interests were at considerable distance from the river itself. The same principle was applied to aquifers, and in particular to the massive Ogallala aquifer in America's Midwest. Ogallala was to be the guarantee that there would never be another dustbowl in the Midwest. As a result the High Plains became the wheat growing heartland not just of the USA, but of the world economy, its 45.5 million cultivated acres producing 10 per cent of the world's total wheat production.[18] It is today in crisis, however, as its water level is historically low.

Local communities, especially Native American collectives, were given tiny allocations of water in the Colorado compact, abandoning the historic community control of the water of the river. Environmental concerns and community needs were ignored. Looking back, we now know that the long-term consequences of dam building have been disastrous on many levels. Since the building of the dam, only 1 per cent of the river water ever reached its delta, once 40 kilometres wide, and since 1960 it has not reached it at all. Mexico has paid the price for the river that simply never reached its estuary. The displacement of local populations as a result of the flooding of large areas for reservoirs has expelled millions of people from their homes and lands – probably close to 80 million on a world scale since the dam building boom began. Those populations, mainly stable farming communities, became migrant labourers tramping the surface of the earth in search of temporary work, possibly building the next dam. The environmental consequences were not addressed at the time nor for the succeeding 40 years. Instead, international financial institutions, especially the World Bank, encouraged and financed a wave of dam building, to the most obvious benefit of the huge multinational engineering consortia commissioned by the state. It was almost invariably the case that costs overran wildly. But more importantly, the big dams themselves, beginning with the Hoover, were monuments to power and the brash confidence of big capital rather than appropriate communal solutions to the long-term problems of water allocation. The very scale of dam building, and the devastating human and environmental consequences, removed any possibility of local control or intervention. Because they were always, in the first place, temples built to modernity,[19] just as they were in the Black Canyon of Nevada in 1936. And in the developing world they became, as Nehru suggested, monumental symbols of nation-building.

By the late 1970s, the US government itself was beginning to change its mind, in the context of a more general review of the environmental impact of water use. Dam building was stopped on grounds of safety and environmental damage[20] though it has also been suggested that the USA had by then reached the limits of available areas in which to build them. In the rest of the world, however, and despite gathering protests, the following decade witnessed a new phase of dam building mainly in the developing countries. The World Bank, the IMF and development banks were highly committed to dams, and disbursed an estimated $75 billion for dam construction by the end of the 1980s. Between 1947 and 1994, the World Bank received 6,000 loan applications for large-scale water projects – *not one* was refused.[21] Dams were seen as evidence of modernisation and development, not simply because of their monumental scale, but because they supplied energy and irrigation projects which would allow new industries and a new industrial agriculture to grow rapidly and enable new states to become participants (but not equal participants) in the global market.

The massive cost of dam building was largely covered by international loans whose high levels of repayment could often prove a crippling commitment. What began to mobilise major movements, however, was the human cost of dams, the displacement of enormous numbers of people (estimates are between 80 and 100 million worldwide), invariably communities of small farmers or indigenous populations in hitherto neglected places. The most powerful protest movements emerged in India, where a series of huge dam projects were represented by the Narmada Dam complex in Gujarat. Despite the sustained protest movement, the Supreme Court of India allowed the height of the dam to be doubled in a series of decisions in the early 2000s.

Yet an Interim report of a Committee of Experts set up by the Ministry of Environment and Forests counselled the immediate suspension of the project until all environmental requirements had been complied with. Arundhati Roy in her iconic essay, 'The greater common good', points out how few dam projects were preceded by proper environmental impact assessments and how rarely they were assessed after their construction. Clearly, however, the priorities of the Indian government in this emblematic case were very different from those of the majority of ordinary Indians. The Supreme Court's argument was that if the river was simply allowed to flow into the sea, it would make no contribution to national development. Yet by the time they were publishing those decisions the negative effects of dams, and indeed of the blocking and diversion of rivers, were widely available and openly discussed.

In 2000, the World Commission on Dams presented its evaluation of the performance of dams. It was excoriating. In its final report it showed huge cost overruns, billion dollar corruption scandals, a failure to deliver promised quantities of water to cities or to irrigation, dramatic shortfalls in the promised amount of electricity delivered, and worst of all 'dams had universally reduced the fertility of flood plains and "invariably" caused erosion of river banks, coastal deltas and even distant coastlines.'[22] Fred Pearce's careful research has exposed what he calls 'widespread ecological destruction' in the service of political arrogance, corruption and private enrichment. Yet while dam building stopped in the USA, elsewhere it continues at a dizzying pace; 45,000 dams have been built since 1960, 22,500 of them in China, second only – despite massive and sustained resistance – to India.

It should be said that the World Commission's report has not gone uncontested. Asit K. Biswas and Cecilia Tortajada of the Third World Centre for Water Management have a long and fierce critique of the anti-dam lobby, accusing it of arrogance, paternalism and a lack of awareness of local conditions in which dams might be an appropriate answer.[23] The response of the anti-dam lobby might be that the major criticism of dam construction is that local conditions, experience and systems of production based on a historic knowledge of climate, soil, crop rotation and the like are the first casualty of dams, a point reinforced by the almost universal absence of proper impact studies before they are built. Biswas himself notes, in the same article, that:

An important question that needs to be asked is why, in the twenty-first century, with major advances in science and technology, it has not been possible to answer the relatively simple question of the real costs and benefits of large dams, so that their net impacts and benefits can be determined authoritatively and comprehensively?[24]

And Pearce adds:

It is too easy to see communities that depend on natural wild resources and the vagaries of untamed rivers as somehow left behind by progress.

The truth, quite often, is the opposite. It is they who have unlocked the truth about how to make the maximum use of natural resources. It is the urban sophisticates with their engineering diplomas who haven't got a clue. The pilot studies on the Logone and the similar research done on the Hadejia-Nguru wetland, cement the case for a completely different way of managing flood plains and the rivers that sustain them.[25]

This one case can be repeated many times over, as dams over time have shown that a significant proportion of their volume at the dam head is lost to evaporation, while the accumulation of dying and rotting plants produced methane in significant quantities, contributing increasingly to global warming.

The build-up of silt at the lake bed reduces the volume of water in the lakes and the levels of oxygen in the water, allowing poisonous algae to accumulate. And the decision by the US government to reduce the number of dams in the country, leaving 5,500 large dams and possibly 50,000 smaller ones,[26] even allowing for the fact that they might have lived out their usefulness, is also an implicit recognition of their negative environmental impact. In 2010, for example, the water levels of Lake Mead behind the Hoover were at their lowest ever.[27] The longer-term effects of salinity and waterlogging change the surrounding wetland and flood plain ecologies forever, negating the stated purposes for which the dams were built in the first place. It may be that there are cases where dams are appropriate, but that has to be demonstrated scientifically to the affected communities. This is not an appeal to leave nature alone, to hold back human development, but it is to say that reckless, short-term, unplanned development, uninformed by an environmental awareness leads inexorably over time to a loss of resources and a diminution of the water resources essential to every phase of development and growth.

There is also the human cost, as we have mentioned; the 80 million plus people who have been displaced as a result of dam building.[28] These are mainly members of farming communities that have existed for many generations on the irrigated flood plains or wetlands. They are not simply expelled from their land; their activities cease, the ecological balance of which they were as much a part as the rivers where they fished or the land they farmed is disturbed for the long term, or for ever, and these people become part of that growing multitude of displaced people, wage labourers or inhabitants of the precarious hidden economy of the expanding cities of the third world. The social consequences are profound and far-reaching; and in the city slums and barrios where they will finally cease their wandering they will, in many cases, pay exorbitant amounts for the water the dams produce. More likely still, they will buy unreliable or contaminated water from street sellers and fall sick from the inadequate sanitation characteristic of these areas, victims of the 80 per cent of diseases that are water-borne.

Since the 1950s, and more intensively since the 1970s, dam building, which was slowing in the USA and Europe, was accelerating in the developing countries. Just as the Hoover Dam was emblematic of a key phase in the USA's dam building programme, Egypt's Aswan Dam was iconic in a third world context demonstrating to a wider world, and to the local populations, the dynamism and power of the state and the commitment to a *particular vision of development*. Today, the largest number of dams are in India and China; 50 per cent are in China, with devastating effects at every level.

China: a case in point

China today has one-fifth of the world population (2.3 billion) which is set to rise to 2.6 billion, but currently uses 14 per cent of the world's freshwater resources. This is less than might have been expected.[29] China has certainly been more efficient in the use of water in agriculture thus far, according to Allan, yet this seems incompatible with its current status as the world's largest manufacturer of pesticides, which it also exports in growing volumes. And the pollution from its accelerating industrial expansion has reached severe dimensions. Its virtual water imports are equivalent to 60 cubic kilometres for food and 14 cubic kilometres for industrial goods. Since the fall of the Soviet Union, China has become a major power in both Africa and Latin America, almost certainly in the first instance because of its urgent need for natural resources as well as markets. In Latin America it plays an increasingly important economic role; in Brazil, Bolivia, Nicaragua and in Venezuela, where the $50 billion 'Chinese Fund' is a major contributor to the Venezuelan budget, in return for which it receives nearly 500,000 barrels of oil per day, a volume which will rise in coming years.[30] Its industrial products – mobile phones, cars, domestic machinery

– are replacing those which Venezuela is no longer able to import from elsewhere. Ecuador has received at least $3 billion in loans to the state, and a number of the mining companies with recently granted concessions there have recently passed from Canadian to Chinese control. China is now the main trading partner for Brazil, with a defining role in its economic and environmental future. Its greatest weight, however, is in financial provision, where it seems to be occupying spaces left behind by the reluctance of the World Bank and the IMF to lend to radical regimes like those in Latin America.[31] Allan's assertion that it can enter into a symbiotic relationship with the living world, however, is sharply contradicted by the state of its rivers, its aquifers and its cities.

Most water-bearing areas are in the south of the country, the source of much of the country's food production. Its industries are mainly in the north on and around the Yellow River Plain, though this region also produces and exports food to the south. The Three Gorges Dam, on the Yangtze River, is the world's largest hydropower station, producing 25,000 megawatts (ten times the power generated by the Hoover Dam); it was also built with an eye to controlling the severe flooding of the lower reaches of the river. China is currently building or has built half the dams in the world, some 22,500. The aim is to reach a capacity of 125,000 megawatts of hydropower, with dams that the government claims avoid pollution, anticipate future climate change and control floods and droughts. Charlton Lewis[32] questions the truth of that reassurance; the dam construction programme, he argues, increases the risk of earthquakes and landslides, destroys precious environments and threatens millions of lives (Three Gorges, we know, has displaced an estimated 1.3 million people). Since the majority of China's rivers originate on the Tibetan plateau and flow rapidly down deep canyons towards India, Pakistan, Bangladesh, Laos, Vietnam, Thailand and Myanmar, there is a high risk of earthquakes as reservoirs rise and fall. Probe International reported in 2012 that around half of China's dams were in areas of potential seismic activity, a fact made all the more dangerous because many of China's dams are cascade dams in succession along the rivers. When one dam fails, the others are likely to collapse under the impact of the emptying reservoirs causing massive damage. The Zipingpu Dam on the Min River was built in 2001, despite warnings of seismic risk from engineers. In 2008 an earthquake just 5 kilometres downstream caused 80,000 deaths.

Figure 3.1   The world's largest dam; the Three Gorges Dam.
Commons Wikipedia.

The claim by the Chinese government that dams release no carbon dioxide, as compared with the coal-fired power stations that provide the bulk of China's energy today, is a sleight of hand. It is well established that rotting vegetation in reservoirs at the head of dams releases methane, which is a greenhouse gas. Neither is the guarantee of limiting floods and droughts particularly credible, since dams interrupt the natural flow of rivers and deprive downstream waterways, lakes and wetlands, which are then subject to drought. Over the century, 50 per cent of the planet's wetlands have disappeared.[33] Reservoirs also accumulate detritus, pesticide contamination and human waste, as well as the silt which traps nutrients which no longer flow down river – allowing sea water to encroach from the estuaries (as has happened with the Indus, for examples, which dries out 80 miles from the sea). China's refusal to sign the Helsinki Rules on the Uses of the Waters of International Rivers agreement of 1966, or the Berlin Rules on Water Resources of 2004, which superseded it, is a direct threat to the downstream nations, especially given China's declared intention to build a dam with 38,000 megawatt capacity on the Yarlung Tsangpo River in Tibet, which later becomes the Brahmaputra and Yamuna rivers in India and Bangladesh. Thirty per cent of China's water therefore comes from Tibet,[34] including its bottled water.

Water scarcity is a perennial problem in China, with its history of devastating floods and droughts. Eight of its 28 provinces are deserts. And though its population is one-fifth of the world total, it contains only one-quarter of the world's water. Its commitment to dam building is first and foremost a question of energy; currently 67 per cent comes from coal-fired power stations, while hydropower covers only 6 per cent of its needs. Although it has backed significant research into alternatives, this cannot in the foreseeable future resolve its energy requirements. The commitment to hydropower, and thus to dam building is clearly therefore irredeemable. The South–North Water Transfer Project, built at a cost of $60 billion, will take water from the south to the industrial developments in the northern plains. But the project embraces a paradox, given that the northern regions produced crops in 2000 with a virtual water content of 26 cubic kilometres, while the Transfer will eventually supply 50 cubic kilometres to the region.[35] It gives some credence to Hook and Pearce's suggestion that this is essentially a vanity project, a huge white elephant or what US commentators have called a 'boondoggle' – which is to say an enormous waste of public money with damaging environmental consequences rarely discussed before they are undertaken. What is certain is that for the moment other energy sources, principally oil, will come from bilateral agreement with countries in Latin America and Africa. For the moment, the scarcity of water restrains the water-intensive alternative of fracking and nuclear energy. But this does not resolve the problem, given the extraordinarily high level of river pollution – 39 per cent of China's rivers are already too toxic for human use, principally because of industrial pollution. And the rapid depletion of its underground aquifers has meant that a number of its major cities are sinking.

The Yellow River, at 5,000 kilometres long, is the fifth longest river in the world and sustains half a billion people with its water.[36] Irrigation using its water has made China a major producer of wheat and maize. In 1938 the deliberate destruction of its major dyke in order to hold back Japanese invaders led to one of the century's great disasters, flooding the North China plain and drowning nearly a million people, not to mention the millions driven from their homes. Yet by the 1970s the river was running dry, failing to reach its estuary for part of every year. Along its courses lakes are disappearing, irrigation channels drying up, and whole areas turning to desert. Yet much of its water remains in its dry upper reaches rather than reaching the fertile plains downriver. En route water is abstracted for mines and for the burgeoning cities, and much of the land on the fertile plain has now been abandoned for lack of water. China is no longer a leading grower of wheat and maize; it now imports both with a major but as yet uncalculated impact on both grain prices and grain stocks worldwide. As Pearce argues, in looking at the problems of silt accumulation along the river and its dialectical history of drought and flood, taming the Yellow River will be a task like Sisyphus's, pushing endlessly against the edges of disaster.

In a comprehensive review of China's water situation in 2009, Peter Gleick, of the Pacific Institute, possibly the world's leading expert on water, argued:

> When water resources are limited or contaminated, or where economic activity is unconstrained and inadequately regulated, serious social problems can arise. In China, these factors have come together in a way that is leading to more severe and complex water challenges than in almost any other place on the planet.
>
> China's water resources are over-allocated, inefficiently used and grossly polluted by human and industrial wastes, to the point that vast stretches of rivers are dead and dying, lakes are cesspools of waste, groundwater aquifers are over-pumped and unsustainably consumed, uncounted species of aquatic life have been driven to extinction and direct adverse impacts on both human and ecosystem health are widespread and growing.[37]

It is a withering verdict on the consequences of unrestrained industrial growth over a period of 20 years and its impact on the environment. Sixteen out of the 20 most polluted cities in the world are in China. And while 88.8 per cent of China's urban population have access to clean drinking water and 70 per cent have adequate sanitation, the rural figures are much lower. Official government estimates are that 300 million rural inhabitants do not have access to safe drinking water. The situation in public health is equally disturbing with deaths from diarrhoeal diseases at 108 per 100,000.[38] As in India, the incidence of cancer directly related to industrial pollution is significantly higher than elsewhere. The specific situation of water, both surface water and aquifers, is critical, with disappearing lakes and wetlands, desertification, and the large-scale pollution of water sources by pesticides and industrial waste. The problem is exacerbated by inefficiency at all levels of administration coupled with corruption and bureaucracy and immobility in central government.

The catalogue of accidents, seepages, pollution of rivers, and the failure to apply any environmental standards at all added to government secrecy, has allowed this potential disaster to develop. The response to these repeated crises has been to seek solutions in massive infrastructural projects, each of which brings its own dangers and negative environmental impacts. The 'dam boom' which has given China half of the dams built in the world since 1950 has produced exactly the same problems that a previous generation of dam builders in the United States discovered, to their cost. But that does not seem to have influenced Chinese government decisions at all. What is clear is the urgency of finding solutions. Unfortunately those alighted on by the Chinese state have included recourse to the private corporations, like Veolia, who have been centrally responsible for the reorientation of water issues towards the market, and the arguments that water provision is subject to its laws and regulations, that have contributed so much to the current water crises. In the twenty-first century, and especially where water flows where it will and respects no national frontiers, the solutions we seek must be global but not globalised. The global capitalist market, in which China is a major participant, can only compete for an increasingly scarce resource. The task of water activists is to remove those considerations from the equation and seek urgent ways to restore and regenerate water, which remains, if properly managed, a renewable resource. There is clearly a growing awareness among the Chinese people of the devastation threatened by dam building; 100,000 demonstrated against the Pubuguo Dam in 2005, until the protests were violently dispersed by the police, and one of its leaders executed.

China is now undertaking dam construction elsewhere, and particularly in Africa, despite the real contradictions that have attended the Chinese experience. The reasons, internally and externally, have a great deal to do with the enormous profits to be made from the construction itself. At least one of the 'Six Companies' who built the Hoover Dam, Bechtel, has gone on to be a major global engineering force; there is ample evidence of the huge amounts involved in dam building, and of the proportion of those costs represented by corruption of public officials.[39] Because in addition to their environmental effects, these massive dam projects are by their nature centralising; their scale strengthens the state bureaucracy and the level of investments arises from their anticipated contribution to the growth of that sector of the economy that will provide returns to merit such enormous expense. The Aswan was above all an instrument to bring Egypt into the global economy; in the case of China, that is beyond doubt.

Vandana Shiva,[40] the leading Indian environmental activist, stresses that water, seen as a commons, belongs to the community as a whole. It is a resource. The dam building business, and its financiers, regard water as a raw material – like coal or diamonds or copper – which is inert until it becomes a generator of cash, until it is bought and sold, in a word until it becomes a commodity. Since the mega-dams we have been discussing are state-led projects, the investors treat the state bureaucrats as owners and disposers of water, and engage them in the commercialisation of the water. Vandana Shiva describes this, in our view accurately, as 'the privatisation of the state' – in other words the process whereby decisions about the disposition of shared resources are made by and for the market, without reference to those to whom it belongs in perpetuity – the community.[41] In many cases, of course, it is not the actual water that is purchased, but land – with the implicit assumption that land ownership automatically carries with it ownership of what lies beneath. This was the principle on which the allocation of water in the USA, be it the Colorado River or the Ogallala aquifer, was determined – and in both those cases the depletion of the resource may have served its investors well, but it has undermined the social interest.

The Indian writer, Arundhati Roy, in her essay, 'The greater common good', argues that:

> Big Dams are to a Nation's 'Development' what Nuclear Bombs are to its Military Arsenal. They are both weapons of mass destruction. They're both weapons Governments use to control their own people. Both Twentieth Century emblems that mark a point in time when human intelligence has outstripped its own instinct for survival.[42]

What is destroyed is a complex farming ecology, replaced by cash crops for export, the only justification for the huge outlay involved in these dams. The massive hydroelectrical output and the availability of high volumes of water, has drawn heavy industrial plants to the region. But the environmental and human impact of the dam, which reproduce all the previous experiences, are offset by the profits to be made (according to the Indian

Supreme Court's final judgement on the matter) – which will, of course, never reach those displaced, contaminated and forced to labour. 'The disputes over dams are, [in the end], a struggle between displaced citizens and the ruthless state machinery.'[43]

The reality is that even the World Bank has cooled on the question of dams. Their negative effects and the protests their construction invariably generates may be one reason. But dams continue to be built, despite the evidence; today it is China that is financing dams abroad, for reasons of profit and for geopolitical reasons too. This is not the result of any change of heart on the part of international financiers but of the dramatic paradox to which we referred in Chapter 2 – that the threat of water scarcity coincides, and not accidentally, with an ideological shift in which water becomes simply one more commodity to be traded in the global markets.

A river runs through it

The Indian state of Kerala had never experienced water scarcity until Coca-Cola was granted a water concession. It pumped groundwater at a rate of 1.5 million litres day, according to Vandana Shiva,[44] drying out three lakes and even some rivers. The plant was one of 67 bottling plants owned by the giant multinational, each of which has provoked fierce local opposition from small farmers forced, they say, to dig ever deeper into local aquifers to irrigate their fields. In many cases, too, Coca-Cola has been ordered to close its plants because of excessive levels of pesticides in the drink. The Coca-Cola situation has become a familiar *cause célèbre* because of the tragic suicides of farmers whose crops were failing; they often cited lack of water as a direct cause of their failure. There is no doubt that Coca-Cola was doing what it has done across the world; diverting water from public supplies or from local rivers and aquifers. Yet the rate of suicides was also rising alarmingly through the first decade of the twenty-first century in several other states.[45]

In the case of states with a dominant export-linked economy, and especially those growing cotton on an industrial scale, small farmers attempting to enter the cash crop business often contracted large debts to purchase herbicides and pesticides, which they could not repay. Since they had also converted from food production, and often from multi-crop farming, there was no fall-back as the moneylenders moved in behind the multinational seed and chemical corporations.

Fred Pearce quotes the example of Coca-Cola with a degree of scepticism; he estimates the Kerala plant's usage as half a million cubic metres,[46] yet even if the higher figure is right, it represents little more than the usage of three rice-growing farmers on a ten-acre property. It does not exempt Coca-Cola, but it does underline the wider origin of the rapid depletion of India's water. India's rapid industrialisation has both drawn on its rivers and polluted them in producing goods, agricultural and industrial, almost entirely destined for markets overseas. Textile factories, like the one Pearce visited on the Noyyal River, pour dyeing and bleaching chemicals directly into the river which then flows to a nearby reservoir, polluting both river and reservoir. From the unlined reservoir, contaminated water seeps into the underground aquifer, from which the surrounding farms draw their water, pumping deeper and deeper as the level underground declines. The great river systems that flow into India and Pakistan from the Himalayas are increasingly compromised, while the Himalayan watershed is dropping at a rate of 4–6 feet per year.[47] And it is not exclusive to the developing world. The iconic Ogallala aquifer in the United States has sustained a wheat-growing industry that produces 75 per cent of the world grain market, which is already oversupplied. But Ogallala is almost exhausted. The Great Lakes on the US–Canadian border are contaminated by industrial and nuclear waste from the plants on its banks. Petrella reports on the IBM factory in Essonne, France, that pumps 2.7 billion cubic metres of industrial waste into the local river every year.[48]

As wells are sunk deeper and deeper, ironically often in an attempt to avoid surface waters – rivers, ponds and so on – polluted with untreated sewage, they begin to reach the bedrock at the base of the aquifers. In the case of the peoples of Bangladesh the decision has lethal consequences, because of the naturally occurring arsenic and fluorides in the water.[49] 'Today, tens of thousands of Bangladeshis have already developed skin lesions, cancers and other symptoms.'[50] The irony is that many of those wells were sunk by international aid agencies and NGOs who failed to test for arsenic and fluorosis. Yet Pearce estimates that in India, China and Pakistan, pumping exceeds recharge by between 150 and 200 cubic kilometres a year. And the problem is not limited to these three countries, though they may be pumping the largest volume of underground water.

In fact, the major consumers of water are the cities of the developing world. Unlike the emerging cities of Europe and the United States, where water and sanitation systems were linked together and planned, the explosive growth of the new cities of the third world was unplanned and chaotic. The self-building of slum cities on waste or unused ground was at one level a testament to the ingenuity of the unnamed popular builders, but at another it meant that services were provided in haphazard and improvised ways by urban authorities largely overwhelmed by the pace of that growth. In what was essentially a succession of emergencies, drinking water was provided as a first necessity, while sanitation came a poor and late second. Indeed the conditions of third world urban expansion meant that amenities of any kind were few, and largely limited to the middle-class areas. There was little or no provision for waste disposal, still less for the disposal of human waste which tended then to leech into local streams or ponds and to return to their river sources, polluting the streams and finding its way into the water tables which were the source of drinking water. It is clear that the larger problem of sanitation cannot be solved except in conjunction with the provision of drinking water through a public supply. But the costs are enormous and the national and international financial agencies have demonstrated their unwillingness to invest in public services that will not produce a profit in the short or medium term, or indeed at all. Thus many of the poor urban populations buy their water from tankers or private water sellers; not only is it often of dubious quality, but its price is many times higher than municipally supplied water. While anyone spending over 3 per cent of their income is regarded as living in water poverty, people in poor districts may pay up to 25 per cent of their income for water, as they do for example in Lima, Port-au-Prince and Lagos.

Meanwhile the effects multiply. The water table in Beijing, for example, has fallen by some 37 metres over 40 years. Other cities too are sinking as the water tables beneath them are mined and not recharged. Las Vegas is one, while Mexico City, built on a drained lake bed is visibly sinking at a rate of 50 centimetres a year. In coastal cities like Jakarta and Bangkok, the lowering of the water table has allowed sea water to intrude several kilometres into the water network. Amman, Delhi and Santiago de Chile, like Mexico,[51] are drawing in water from further and further away – in Mexico City's case from well over 300 kilometres distance.

But an equally urgent problem in cities is what is euphemistically described as 'unaccounted-for water'. This may refer to water that is diverted, or not to put too fine a point on it, stolen from the public water supply, but more significantly it refers to water lost through leakage. The scale of the problem is astonishingly high; 17 per cent in the USA, 39 per cent in Africa, 42 per cent in Asia and Latin America, and 52 per cent in the cities of North Africa and the Middle East.[52] In Manila and Johannesburg it is above 50 per cent, 58 per cent in Manila. To put the figure in context 50 per cent represents something like 1.5 million cubic metres per day. And this is not only the case in developing countries; in Boston leaks rise to 30 per cent, in London 50 per cent.[53] There are counter-examples which are more heartening – Singapore, for example, where the figure is 8 per cent and Tokyo, the lowest at 4 per cent. The last two examples make it clear that leaks and the major loss of water that they represent, are the result of the failure to renew pipes and infrastructure and the reluctance to invest in public services which have become all the more common since privatisation and neo-liberalism brought with them systematic cutbacks in public service budgets. The transfer of responsibility for water provision into the hands of private companies brought no improvement; there is, after all, no profit to be made from renewing infrastructure. It appears in the account books of Suez and the rest as a cost, and one they are not willing to bear; it is the distribution and supply of water that produces revenue. Therefore the responsibility for maintaining the pipes and pumps falls back on cash-strapped municipalities.

The problem is doubled when we consider sanitation, sewage and human waste disposal. Open sewers are still a common sight in hillside slums and there is rarely any kind of oversight of waste disposal. And just as the water pipes are prone to leaks that seep into the subsoil, so too are the channels carrying raw sewage. Thus water returning to rivers, lakes or aquifers from urban concentrations is often polluted. Cholera becomes emblematic of the deteriorating health conditions of many third world city dwellers; the 1990 outbreak in Peru, which spread through the southern part of Latin America, affected 400,000 people. The KwaZulu outbreak of 2000 in South Africa left 265 dead. In 1994, 500,000 people in the Indian city of Sutar fled an outbreak of pneumonic plague that spread from rotting cattle carcasses in the river. In Latin America, 98 per cent of domestic sewage is discharged into streams. It is certainly the case in Mexico.

Worldwide 10 million people die from water-borne diseases, according to the UN.[54] The translation of the Western model of sanitation, water mains and sewage pipes, was simply inappropriate for the poor populations of these cities, yet it was the one held to by the majority of city administrators – which in practice meant, in most cases, that nothing was done. As Maggie Black persuasively argues, it is the people at the bottom, the poor and the exploited, who suffer 80 per cent of the water-borne diseases defined by the WHO, and whose children are most likely to die from the diarrhoeal range of diseases like cholera and typhoid.[55] Yet the provision of sanitation, despite being linked with the provision of drinking water in the Millennium Development Goals, has fallen badly by the wayside. And even where it is provided, it is insufficient of itself unless there is water for washing too. Hygiene is a broader issue than the provision of sewage facilities because it concerns social custom, tradition and usage.[56] And as she points out that is critically about involvement and control; clean water alone will not solve the problem. Indeed there are many circumstances where sewage disposal by water is costly and wasteful; there are a number of alternatives to which we shall return in later chapters.

In their 2012 report, the Blacksmith Institute and Green Cross looks at the wide variety of dangerous and toxic waste materials to which populations in the developing world are vulnerable. The 2 million tons of radioactive waste on the banks of the Mailuu-Suu River in Kyrgyzstan must figure among the most shocking, but the most dangerous site they identify, the Agbogbloshie dump in Ghana, with its 450,000 tons of poisonous e-waste – that is discarded electronic materials and equipment – has a particular poignancy, since it represents the results of an attempt by Ghana to develop a computer and mobile phone industry of their own. This attempt to open a space in the global markets has only produced the highest toxic threat to life on the planet, surpassing even Chernobyl.[57] It is not a surprise to learn that 25 per cent of deaths in the developing world are the result of one form or another of environmental pollution. But the report provides an explanation, derived from a HO/UNEP report which found that 'barriers to addressing environmental pollution are economic, institutional, political and social in nature and include trade globalization, market liberalization, debt burdens and structural adjustment policies.'[58]

And the final bitter irony is, as the report shows, that Europe and North America have, especially in the last two decades, taken increasingly stringent measures to control pollution of water and urban environments, while at the same time exporting their most polluting industries to protect their own environment at the expense of others.

The water tragedies, of which we have mentioned just a tiny proportion, multiply year on year with devastating consequences. Where water is a weapon, where water is diverted and the ecosystem is fatally disrupted, where water becomes a commodity, the consequences are the same. The death of thousands, the impoverishment of millions, the deepening of inequality.

4

A Short Trip Through Amazonia

THE LAST JOURNEY OF THE GOLDFISH

The Amazon rises at the Nevado de Mismi near the border between Peru and Bolivia and 100 kilometres from Lake Titicaca, in the cold altiplano of the High Andes. The goldfish that is one of the Amazon's less obvious exports begins its journey in the cold high lakes.

From there it might travel to Ucayali in the Peruvian Amazon, not far from the Andean redoubts of the Inca Empire and the mountain universe the Incas called Tawantinsuyu. Madre de Dios brings him to the border with Brazil, and here the pure cold flow begins to acquire a human population, miners using high-pressure hoses to gouge pits out of the river bank. There are several of these mines in Peru, this one owned jointly by Buenaventura and Newmont, the second largest gold mining company in the world. Yanacocha, in the mountains of Cajamarca province, has been the scene of strikes and protests for several years. Flowing down the Ucayali towards the Marañón, the water begins to change colour as the pollutants from the mining and oil operations run from the banks into the river.

In the Pastaza Morona region of Ecuador, acknowledged as one of the most remarkable ecosystems in the world, everything is stained black, thick sludge from the oil and mining operations flows into the territory of the Shuar people. Kinross, Talisman Oil, Ivanhoe, Tripetrol extract oil and minerals here. The people who live here have no water, nor baths to wash off the slime. They have to buy their water from private tanker lorries; meanwhile thousands of gallons pour hourly into the river from the mine workings.

In the Putumayo and Sucumbíos in Colombia, the oil and mining concessions continue. The local people have protested that their water is contaminated by the mines, or by the pesticides used by the flower growers.

▶

Once the area was closed to foreign companies and controlled by guerrillas and drug traffickers. With US support through Plan Colombia, the region is largely pacified – and the foreign companies have moved in, looking for oil, cutting back the forests, turning rivers into Coca-Cola.

Further down the river huge boats carry the logo 'Maggi' – Brazil's largest producer of soya. Other logos begin to appear, like Coke and Pepsi Cola; they take their water from here, though they claim to filter it and put most of it back. The Marañón, the Ucayali and the great Rio Negro meet at Manaus, a major city, noisy, dirty, pressured, where the rivers become the Amazon. Sony and Panasonic both have plants here. Down towards the Xingu a giant dam, the Belo Monte is being built after a 20-year delay because of the resistance of the Kayapo people and their Chief, Raoni. Not far away are the aluminium smelters owned by two Chinese companies; they will use most of the water from Belo Monte, the largest dam in the world after China's Three Gorges. From there it is a short swim to the sea at the port of Santarém, where Cargill, the world's largest private agricultural corporation, loads its soya for Europe and China.

Amazonia holds 20 per cent of the world's fresh water; the statistic is well known and often repeated, though for reasons that will become clear that figure is being revised downward – some writers now suggest it is 12 per cent. We will speak of Amazonia rather than of the River Amazon, which in fact is just one – although the widest (between one and six miles across) and the second longest – of the 200 rivers flowing through this giant watershed. It extends into the territory of eight Latin American countries: Peru, Ecuador, Bolivia, Colombia, Venezuela, Guyana, Suriname and Brazil, which contains the largest segment of Amazonia. This massive rainforest region, often described as the world's lung, extends for 2.3 million square miles, its vegetation absorbing carbon dioxide and exhaling 20 per cent of the planet's oxygen. Where the Amazon meets the ocean, 55 million gallons of fresh water flow into the Atlantic every second, one-fifth of all the water discharged by rivers on earth. In its trees, foliage, and rivers live two-fifths of all the fauna and flora species on the planet. But it is not a virgin forest. It has been home to human beings since time immemorial, their social and cultural diversity is as important an aspect of our present and future as its biodiversity.

These statistics are probably familiar to everyone; and it would be easy to absorb them into a general discussion about the casualties of progress and the need for conservation, as if what was needed were museum displays of humanity's past – a record of human history. That is one story, much favoured in the West. But Amazonia represents something very different – a narrative of wanton destruction whose effects can only be guessed at, but which evidences the short-term thinking that characterises capitalism as a system. The long-term consequences have rarely been considered, let alone informed economic or political strategies. Until now, the critique of the wilful destruction of the Amazon has been couched in terms of a romantic vision of a primitive world under threat, and an equally sentimental Western adoption of indigenous concepts like 'buen vivir' (the good life) which would have much to offer if they were translated into a global context. But the global captains of industry have worked hard to consign that discussion to a nostalgic irrationalism.[1]

The Amazon has been a magnet for seekers after wealth since it first entered the Western consciousness. Gold, cacao, cinnamon and clove, vanilla, Brazil nuts, Brazil wood, jute and, at the beginning of the twentieth century rubber have attracted the world's attention and stimulated its greed through time.[2]

Each of these intrusions wreaked their own kind of havoc. But the late twentieth century has accelerated the destruction of the rainforest to a degree that even Henry Ford might have found surprising. The products that the Amazon gives to the global market as the twenty-first century begins are soya beans, cattle, oil and gas, bauxite, manganese, iron, kaolin, aluminium, copper, coltan – and goldfish. All but the last require the destruction of this precious resource. Here the figures are truly alarming. Already 30 per cent of the rainforest has been cleared – that is, around 17,500 square kilometres per year (an area roughly the size of Wales) which amounts to the total area of France and Germany combined thus far – environmental destruction on a massive scale. As the list above shows, much of the new activity in the Amazon region is extractive – mining and oil drilling. Both these industries employ vast amounts of water, which is returned to the rivers full of contaminants, chemicals and minerals. The minerals, however, are transported elsewhere – increasingly these days, to China. Thus pollution is added to the decimation of the forest. The irony is that the forest has been cut down not just for its wood but to clear land for cattle and soya production; the precious woods are transported to China and transformed into elegant furniture for the wealthy, while the rest is reduced to charcoal used mainly by the steel industry, itself a major polluter of the watershed. It is, by any calculation, a very unequal exchange, from which neither the local populations nor indeed the environment itself, derive any benefit (other than starvation wages).

All of these processes use huge amounts of water. Much of it is contaminated; much more is exported in virtual form, embedded in the maize and soya exported for bio-ethanol, in the meat exported across the world, in the minerals taken to China and the developed world to sustain a new industrial boom.

Brazil itself, committed to an economic development which will ensure its place as seventh among the world's richest nations and thus exempted from a number of environmental restraints, is once again intensifying the rasing of the forest under Dilma Rousseff, having held a moratorium for a period under the previous presidency of Lula.

The symbol of the times in the Brazilian Amazon is the Belo Monte Dam, on the Xingu River, to be completed in 2015 at a cost of $19 billion, which will produce an estimated 11,233 megawatts of electricity by this year. Its economic cost is only part of the price paid for electricity supplies largely destined for the aluminium plants in a regional industrial zone: 500 square kilometres have been flooded, and tens of thousands of local indigenous families displaced. And it is only the largest of 600 dams in the Brazilian Amazon, supplying the hydroelectricity, which represents 60 per cent of the country's energy consumption. In this sense it is 'green' energy, that is, an alternative to fossil fuels, but as is the case with all dams, with major environmental effects – changing the oxygen content of rivers which will destroy the fish population, forced displacement of local populations, large-scale immigration of workers which will bring its own demands on the environment.[3] According to Philip Fearnside, Belo Monte 'has also functioned as a "spearhead" in creating precedents that weaken Brazil's environmental licensing system'.[4]

More controversial still is the Madeira River Hydroelectric Complex, including the Santo Antônio and Jirau Dams, which will make the Madeira River navigable into Peru and Bolivia. It is largely being built with Chinese machinery and for the benefit of exports to China. But in addition the extension of carbon credits for hydroelectricity projects has lowered the cost, and raised the profitability of dam construction.

Simun Farabundo[5] sees this as a second wave of external exploitation akin to the process described by Eduardo Galeano in his iconic study *The Open Veins of Latin America*.[6] As we have noted above, Farabundo's conclusions are more than confirmed by the level of new extractive activity throughout the region. It is the scale of dam-building, and the purposes to which the energy produced is dedicated that is the issue here. The São Paulo case, which we discuss below, suggests clearly that it is industry and export agriculture who are the majority users of that energy, while water becomes scarcer for the human consumer. The purpose of the Madeira Dam is to facilitate the export of grains (particularly soya), timber and minerals to the port of Belém and thence to Europe and China. The additional 35 million tons of soya beans that will be transported will be grown on land clawed back from the rainforest, increasing deforestation (by 600% according to some reports) and the negative climate effects that this invariably produces.

The completion of the Madeira Dam will open what will be a major shipping lane across the north of Amazonia; a similar proposed channel in the Paraná River (the Hidrovia, currently suspended after years of argument) will create an outlet for the soya and cattle produced in the south of the region to be exported directly.

Brazil has a record of active involvement in the implementation of environmental regulations of which, paradoxically, the intensive development of hydroelectric power is one, contradictory, element. It is contradictory because the move towards zero-emissions energy is wholly positive, but the scale of that development, and in particular in relation to dam building, reflects the aggressive growth of extractive industries and export agriculture whose lion's share of energy consumption distorts policy considerations, particularly when the levels of democratic public consultation over future planning are very low. It would appear that the demands from industry and people are in direct conflict. In the Paraná, however, there is an emerging project for community participation and ecological balance.

The picture becomes more disturbing and action more urgent when we pull back the camera to include the Colombian, Peruvian, Bolivian and Ecuadorean Amazon regions. Now 5 per cent of the Peruvian Amazon is given over to oil and gas exploration, and it is feasible that by 2020 that figure will rise to an astronomical 75 per cent.[7] Coca cultivation in both Peru and Bolivia has risen exponentially,[8] just as it has declined in the Colombian Amazon, in part at least because of the sustained repression of cocaine plantings. The preparation of cocaine base largely involves kerosene, acetone and solvents later pumped into rivers to join the mercury from expanding gold mining. As gas and oil production have increased, the level of indigenous resistance has risen correspondingly. Since 2008 there have been 12,000 such protests, echoing the bitter struggles against the mining concessions across the Andean region. None of the multinationals ploughing their way into the area have had the slightest compunction in responding with extreme violence, using their own private security forces. The indigenous people, for their part, have enjoyed little or no protection from states deeply embroiled with the companies. In Ecuador, the Yasuni in the Ecuadorean Amazon has become emblematic in conservation circles; it is a precious and unusual biosphere, for very specific reasons.[9] An international campaign involving Hollywood stars and prominent celebrities has campaigned for its protection, especially against the background of the level of damage wreaked by Chevron-Texaco in its earlier exploitation of the region. The Ecuadorean Supreme Court calculated that damage at $9 billion, a case fought and still being fought in US courts whose sympathies are undivided. The response of Ecuador's president Rafael Correa some years ago was to offer the world community the option of conserving the region; if it could collect $100 million to finance alternative projects, the oil would stay in the ground. There were sympathetic noises from some European countries, until the financial crisis of 2008 stopped the discussion in its tracks. Since then the whole of the Ecuadorean Amazon region, with the exception as of now of Yasuni, has been parcelled out in oil and gas concessions mainly to Canadian and Chinese companies. Yet the struggle continues there too, where the Shuar nations and others have organised the defence of their territories and of the water beneath them. The Ecuadorean government's response has been to criminalise their protests.

In Bolivia, a government that takes enormous pride in its representativity and its indigenous heritage faced in 2011 the complex contradictions of a continuing involvement in extractive industry and agro-exports in the TIPNIS national park. There the building of a major highway through the area had little to do with local farming communities; the road would connect Bolivia's eastern regions (Beni and Santa Cruz) with the vast soya plantations and cattle farms on either side of its frontiers with Brazil, many of which are Brazilian owned. In fact the whole project is Brazilian-financed. The government has argued that the road will enhance tourism, while the coca farmers organisations demand the right to expand their cultivation into the region. And it would almost certainly (though no one is admitting it) provide a means of transporting minerals from the Andean mining centres to the waterways that would carry them to Europe and Asia. The communities of TIPNIS initiated a protest march towards La Paz, but were met as they approached the capital by a police blockade. The repressive response provoked a significant movement of solidarity, and the project was suspended. As Vice-President Álvaro García Linera later acknowledged (in 2013), there had been no consultation with local communities and he promised that the project would be suspended indefinitely. In 2014, however, the government of Evo Morales announced the relaunch of the project arguing that there is a national interest that prevails over that of the TIPNIS community.[10] In reality it exemplifies the contradictions inherent in programmes of development based on extractive and agro-export industries controlled by multinational interests and shaped by the global market. Bolivian vice-president Álvaro García Linera has described as a 'creative conflict of interests' this face-off between the global market and its relentless pursuit of profit in the here and now and a project of sustained and sustainable growth which places humans and their environment at the centre of the socio-economic equation. That is the conflict that is active in every area of the Amazon region – and whose outcome will affect everyone who breathes the oxygen that the Amazon so generously provides. Water flows through every facet of this millennial dilemma, as metaphor and as reality.

The year 2015 has already produced the most dramatic illustration of the problem. São Paulo, a megacity of 20 million, where water has been rationed for the past year, is suffering 'water and financial penury'.[11] Water is available for two out of four days, but with little or no prior warning. Prices have risen, while water pressure has been lowered, producing increased contamination and a widespread incidence of dysentery. People have been asked not to take baths. Hospitals are short of water permanently and the city's main dialysis unit, which cannot function without regular supplies of water, has had its supplies cut without explanation and with no timetable. Unsurprisingly the price of bottled water jumped by over 20 per cent between January 2014 and February 2015 (while inflation stood at 7.7%). There is clearly no disaster which cannot represent an opportunity for gain for some! The causes of the shortage, when it was finally acknowledged, was a prolonged drought and the draining of the local reservoir, Cantareira – currently at

10.7 per cent of its capacity. The key issue about this story, however, is what the water authorities see as their priority in similar situations. Despite the clear guidelines set out in Brazil's Lei das Aguas that priority should always be given to human consumption, in São Paulo's case the director of the local water company Sabesp was very clear – the company could not break its contract with 500 local cattle ranching and soya growing companies. Beyond that Brazil exports 112 trillion litres per annum embedded in its exported products.

Currently, and despite expectations, Brazil stands in fourth place in its levels of contamination by greenhouse gases emitted as a result of deforestation. If one of every five breaths we take comes from the Amazon, we would do well to act together to conserve the forest or learn shallow breathing as a matter of urgency.

Amazonia represents 35 per cent of Colombia's national territory. Ironically it was kept relatively untouched until recently because the region was controlled by guerrillas and drug traffickers. In 2012, when the 'pacification' process (under the aegis of Plan Colombia) reached a culmination and talks with the guerrilla groups began, the government began to offer concessions to the corporate interests who had long lusted after the area – for its forests, for the opportunity to extend soybean production in the region (which could then be exported through Brazil) and to explore the oil and mineral deposits the region concealed.

Between 1998 and 2005, just 209 hectares per year were allocated as concessions. Between 2006 and 2012, it was 16,000 hectares per year.[12] The year 2012 saw the highest levels of deforestation in the Amazonian regions outside Brazil – between 2004 and 2012, 541,931 hectares of forest were cut down (42% of them illegally).[13] In the case of Colombia only 20 per cent of the area conceded to oil, gas and mining concessions is under exploration – but a further 40 per cent is being technically assessed in preparation.

Today China is Brazil's largest trading partner, it has extended loans to the other countries of the Amazon region and is actively involved in mining, deforestation, iron and aluminium smelting, the timber industry (although it is forbidden to cut precious woods in the Amazon, their shortage elsewhere suggests that like other prohibitions these too will yield to the pressure of money and neo-liberalism).

The perplexing thing is that while international pressure over the devastation of Amazonia was intense while the recipients of the products of the region were in Europe and America, things have gone very quiet since China became the main driver of the intensified and accelerated exploitation of the planet's lungs. Perhaps it is because China is more parsimonious with its information, perhaps there is some confusion as to the political and economic nature of China. Yet the level and manner of China's exploitation of Latin America's 'open veins', the driving force behind its interventions in the world economy, can surely leave no doubt that this is an aggressive capitalist economy accumulating by dispossession in a manner indistinguishable from the USA and the other 'lords of water'.

5

Bitter Harvests

The Green Revolution: a promise unfulfilled

The 'Green Revolution' began in Mexico where an American scientist called Norman Borlaug developed a new disease-resistant, high yield variety of wheat.

In the 1950s new varieties of maize were added and in the next decade, with support from the Rockefeller and Ford Foundations, his research extended to India where a new variety of rice, IR8, produced more grain per plant. The selective breeding of plants produced larger seeds, and plants that used photosynthesis more successfully and thus could be planted in a variety of climates. It was a development that was widely celebrated in a period in which there was increasing public concern about famine. The word 'famine' is highly charged, evoking the images of mass starvation most recently associated with the Ethiopian famine of 1984, explained by and large as the effect of some natural disaster – desertification, flooding etc. Most historical famines, in fact, have been the result of policy decisions by ruling groups either designed as a kind of collective punishment, like the Indian famines of the mid-nineteenth century or the Vietnamese famine of the mid-1940s, or the result of authoritarian decisions affecting large sectors of the population, as in the Soviet famines of the early 1930s or the famine in China during the Great Leap Forward of 1958–61. Indeed, it was the British Famine Commission of 1880 that concluded that famines were not the result of food scarcity but of a failure to ensure its proper and equitable distribution – a conclusion that has been dramatically confirmed through subsequent history.[1]

The last major famine in India was in Bihar in 1966. But in fact the rising concern about the possibility of famine and widespread food scarcity was more probably a response to neo-Malthusian arguments about population growth which became current in the 1960s.[2] In that sense the Green

Revolution appeared to provide an answer – the possibility of increasing two or three times the productivity of agriculture. Its successor in later times, the concept of 'genetic modification', has received a far less positive press, smacking as it does of manipulation by anonymous corporations, especially the giant in the field Monsanto. Yet they are, in fact, similar processes. The Green Revolution was offered as a solution to the necessity to increase food production for a growing world population. Mexico and India (Punjab), the first areas of application of Borlaug's new techniques, both successfully and significantly increased their output of maize and rice.

Mark Dowie, the campaigning journalist, reflecting back on the Green Revolution, emphasised the support given to it by the major US foundations, Ford and Rockefeller:

> On any list of the greatest grants [...] the Green Revolution would rank in the top five. That ranking – if its social and ecological consequences are ignored – is well deserved. Although hunger has not been ended by the revolution, it has been reduced; enough food is now produced, year after year, to feed the world – with some to spare. However [...] there are still about 800 million hungry people in the world and 185 million seriously malnourished preschool children.

Dowie argues that their patronage of the Green Revolution had a purpose that was far from philanthropic. 'The primary objective of the program was geopolitical: to provide food for the populace in undeveloped countries and so bring social stability and weaken the fomenting of communist insurgency.'[3] And in this purpose, it was largely successful.

In the post-Second World War context of colonial liberation movements and anti-imperialism, hunger was seen as a political rather than a technical issue. It was centrally about the unequal distribution of wealth rather than agricultural methods, and there was increasing agitation for agrarian reform across the developing world. As Dowie points out, its intensified farming methods and development of high crop yields did hold back the spectre of mass hunger – but at an enormous long-term environmental and social cost. While the Green Revolution produced new crop varieties that were more productive and resistant, it did so with the massive use of pesticides and herbicides whose long-term effects have only recently been recognised and catalogued. And it multiplied several times the amount of water used for cultivation. This, together with the large-scale use of farm machinery increased the cost of the Green Revolution; in the short-term small farmers incurred debts they could not pay in purchasing the new chemicals, while Green Revolution methods lent themselves above all to large-scale industrial farming. The longer-term consequence of the Green Revolution was an increasing concentration of agricultural production on those cash crops – like rice and wheat – which were the beneficiaries of the research, and the subsequent reduction in the range of plant varieties. The result was a concentration on grains, rice and animal feed, at the expense of subsistence crops, which tended to be cultivated through polyculture – the cultivation of several different crops corresponding to principles of rotation, and adapted to specific climatic and soil conditions. The wheat-growing areas of the United States, which had imported wheat until the 1940s, became by the 1970s net exporters of grain, supplying some 75 per cent of world demand by the end of the century. The Punjab, where the experimental cultivation of the new rice variety IR8 began, became a major exporter of rice. But in the Philippines, where the new strain was planted widely, fish and frog populations in the paddy fields were dramatically reduced. Similar impacts were noted elsewhere as the diversity of plants was reduced in response to the high yield varieties. Over time, they began to develop resistance to pesticides, as did the bacteria in the soil and some of the insect life around them, requiring their use in greater volumes. In the Punjab, where crop volumes are falling, the long-term effects of the use of pesticides, herbicides and insecticides have included a very significant increase in the incidence of cancer. In what to us seemed a shocking finding, the Blacksmiths Institute's 2013 Report, concludes: 'When sprayed, only 1% of pesticides end up being effectively utilized; in most instances they are distributed into the air and water then run off takes them to surrounding waterways.'[4]

The resilience of these pesticides is evidenced by the fact that DDT has been found in Greenland ice and in the blood of Antarctic penguins. We may assume therefore that the 99 per cent that does not settle on the fields finds its way into human organisms and fish stocks via the food chain and in the water drawn from rivers and lakes. One-third of the world's crops are produced through the use of pesticides and herbicides. The largest manufacturer and exporter of these compounds today is China, whose rivers are now so contaminated that many of them are not fit for human contact. It has therefore replaced Monsanto, the largest producer of GM seeds until recently, whose impact on agriculture and on the life of farmers everywhere has been dramatic. Vandana Shiva[5] notes the origins of the largest pesticide companies, like Monsanto and Dow Chemicals,[6] in the production of early chemical and biological weapons. That history is reflected in some of the names given to pesticides – 'Round-Up', 'Machete', 'Lasso' and 'Squadron', for example.[7]

The evidence of the harmful effects of the chemicals employed increasingly in agriculture is weighty, and with it the growing catalogue of disease and death associated with them. The declining productivity of agricultural land is demonstrable too. But, it might be objected, mankind must eat and given projected population growth, agriculture must grow more intensive and the amount of land devoted to food production must increase. That seems obvious, yet here there is a contradiction in terms. Current methods of large-scale farming will yield diminishing returns, and it is likely that while absolute hunger may decline, malnutrition will increase. The 28 per cent of the world's population living and working on the land are also its poorest sector and the most subject to hunger. The paradox is highly significant; small-scale farming has been driven to the wall by industrialised agriculture, yet it is far more likely to maintain the ecological balance and conserve resources to ensure the future productivity of land. And to emphasise the more general point, the World Resources Institute Report 2013–15[8] points out that although more and more land is dedicated to crop production there are still 870 million undernourished people in the world. Thirty-seven per cent of the land and 70 per cent of the world's freshwater is used for food production, a large proportion through irrigation. Yet the water that runs through the irrigation channels will in many cases contribute to the developing water crisis, by returning polluted water to rivers and aquifers.

The Institute calculates the gap between current food provision and food needs for an estimated 9.6 billion people by 2050 at 70 per cent; but it notes that if biofuel production (which is especially heavy on water use) is not included that figure would reduce by 10 per cent. Wider bioenergy targets would require humanity to at least double the world's annual harvest of plant materials. For that reason, it concludes, 'the quest for bioenergy at a reasonable scale is both unrealistic and unsustainable.'[9]

Between 1990 and 2000 world population rose by 2.6 billion, but food production kept pace with the increase. Yet that tells a very partial story,

Figure 5.1 'Beware the Water Laws', on a wall outside the headquarters of CONAIE, Quito, Ecuador. © Mike Gonzalez

because that food was not equitably distributed so that despite the increase 870 million people are still hungry. Second, the rate of increase slowed through that period and the first decade of the twenty-first century. The reasons are complex but crucially important for an understanding of how best to confront the water crisis in years to come.[10] As agriculture became increasingly industrialised, small-scale and subsistence farming was squeezed out, as we have suggested – by the cost of maintaining production in a system dominated increasingly by expensive chemical pesticides and insecticides, the increasing cost of water, and the growing difficulty of competing with global food producers and distributors many of whom – in the USA for example – received government subsidies.[11] Yet the WTO, representing the big growers, vehemently objected to subsidy regimes in the developing world as unfair competition. Others were driven off their land by dams, reservoirs and rural poverty, and moved towards the mushrooming cities of the developing world. This had two effects. First, the volume of water channelled towards the cities grew exponentially, although as we have seen the proportion of loss, waste and misuse grew at a proportional rate too. Nevertheless this direct water flows away from agricultural land, and the pollution common in those cities in turn affects both surface water (rivers and lakes) and underground water sources. At the same time urban expansion intrudes into agricultural land. Both phenomena disrupt the water cycle. And on the irrigated land itself increased salinity and waterlogging on flood plains served by reservoirs exhausts the land and lowers water productivity still further. The most telling example is the Nile flood plain, where something approaching 15 per cent of reservoir water is lost per year to evaporation and the slowing of river flows leads to the accumulation of rotting vegetation at the dam heads that increases methane emissions. The rising water table, in its turn, limits the oxygen reaching underwater plants generating waterlogging. In a word 'the decline in farmland productivity is closely tied to interventions in the water cycle and ecological disruption caused by Green Revolution technology.'[12] This is not a tendentious position deriving from environmentalist ideology, but a conclusion drawn by the vast bulk of analysts on the basis of the stock of available scientific data.[13] In 1998, 1430 cubic kilometres of water were lost as a result of irrigation; by 2025 that will rise to some 1485 cubic kilometres. But it will need to use methods of irrigation that have learned from the experience of the Green Revolution. It will need to avoid the traps of salinity, the destruction of aquatic ecosystems, and adapt to the decline of freshwater availability in lakes, rivers and aquifers.[14]

Vandana Shiva, the leading Indian scientist and water activist has been relentless in her criticism of Green Revolution technologies, as have many others. Her argument is that the Green Revolution, the so-called 'miracle' of high yield crops, is a myth and a diversion. They should, she argues, be called 'high *response* crops' because they are designed (the word is apposite) to respond to chemicals and to intensive irrigation techniques. The specificity of that relationship between plant and chemical fertilisers excludes other plants and other varieties, and leads inexorably to the elimination of an agriculture of diversity in which there is a balanced relationship between many elements, different plants, bacteria, organic elements in the bio mass – an ecology in other words. Instead, the new dwarf plants, heavy with seeds above ground but shallow-rooted below, contribute little to the soil precisely because they cannot hold water. These crops are increasingly dependent, therefore, on the chemicals that shower down on them, chemicals that make their way into the soil and into the run-off, while the water – filled with con-taminants – runs off into the nearest waterways. As they become resistant to one set of pesticides, new and stronger ones take their place. In the process, ten times the water is employed in their cultivation than in non-Green Revolution crops. And the long-term result is the reduction of agricultural diversity to monocrop cultivation, which are above all cash crops for export.

If, as Gunnar Myrdal suggests,[15] hunger and famine, and the possibility of a mass popular resistance were the real motivations behind the enthusiastic adoption of the Green Revolution, or as Shiva puts it the Green Revolution was designed to head off the 'red revolution', then the massive financial support given by US foundations like Ford and Rockefeller, US government subsidies, and the backing of the World Bank are explained. The miracle of intensified rice cultivation in the Punjab, for example, was short-lived; the over 1,000 varieties of rice that existed, and the 250 crops traditionally grown there, were reduced to just one, IR8 rice. It was represented as a solution to the problem of hunger; yet it undermined the capacity of farmers to produce the variety of food crops they had once grown, and created a dependence on food *imports*, which local farmers could ill afford, especially since they were rapidly indebted to the fertiliser companies whose products were indispensable to the maintenance of the green revolution. Many of them had grown millet, a crop that required very low quantities of water; they were now obliged by the World Bank, as a condition of subsidy, to grow sugar cane and cotton, both of which were water-intensive export crops, as opposed to millet.

It was a clear indication of where the 'Green Revolution' was leading, and what interests it favoured. The promise to feed the hungry proved to be illusory; not only did the number of people suffering hunger remain approximately the same between 1980 and 2000; the numbers suffering from malnutrition also rose – and this despite the fact that 28 million hectares were now growing GM crops. The problem was that the productivity of that land was declining and the available good quality ground water was also decreasing in volume, as the pollution from chemicals permeated rivers and aquifers were depleted faster than recharge. Over time the excess of nutrients in fertilisers causes the plant to absorb excessive amounts of the micronutrients in the soil 'inhibiting the plant's capacity to absorb the nutrients provided by the fertilisers.'[16] In an aquatic environment, this in turn can cause ecological disturbances that are highly disruptive, like algae blooms and the loss of oxygen – or hypoxia – in the water. The consequences can be confirmed, in the most extreme form, in the Gulf of Mexico where the waters of the Mississippi, heavy with fertiliser, disgorge into the Gulf. Effectively it is a massive 'dead zone' in which the fish stocks that provided for Louisiana's important fishing industry are disappearing – an industry that suffered another major blow with the explosion of BP's Deepwater Horizon Platform in the Gulf of Mexico in 2010. Similar effects have been registered elsewhere, in the Adriatic and the Black Sea for example. And there is increasing evidence of extremely serious effects on human health.

'Chronic health problems associated with pesticide exposure include cancer, birth defects and nervous-system damage.'[17] The WHO reports that between 2–5 million people die each year from water-related diseases, most of them children and all of them living in the poorest parts of the developing world. The causes of the rise in cholera and typhoid are, directly, poor sanitation, the discharge of human waste and industrial waste into freshwater sources, and the dissemination of chemicals from agriculture into rivers and aquifers.

In addition many fertilisers include endocrine-disrupters, one result of which is a global fall in male sperm counts. Fish stocks too, the source of 16 per cent of the animal protein consumed worldwide, are declining; and dams have also had a significant part to play in that decline, blocking migration routes, increasing salinity in estuaries and creating hypoxic zones, like the Louisiana shelf. One result, emblematic though affecting a very small minority of the world's population, is that the wild sturgeon providing the world's most expensive caviars are now virtually extinct.

The figures and the horror stories abound, but they can be simply summarised – in Constance Hunt's words, 'pesticides are poisons by design'. And while it has been claimed that what is called the 'second Green Revolution' – in other words genetic engineering – will address and right the errors of the first, everything indicates the opposite. If the first Green Revolution, while claiming to be directed at small farmers worldwide, in fact favoured large landowners – the second has intensified the power of the agricultural monopolies worldwide. While several major companies manufactured the first generation of pesticides, the second generation have been produced increasingly by one monster multinational – Monsanto. Vandana Shiva emphasises, furthermore, that while the first Green Revolution was conducted under some degree of public scrutiny, at least within the USA, the second is a wholly commercial operation, dominated by giants like Monsanto and Cargill. So significant was the opening of these new fields of research into bioengineering that a meeting of scientists in 1972 called for a moratorium on research and development into genetic engineering until the issues, scientific and ethical, were better understood. Their warnings and concerns were ignored. Today, research in molecular biology, once largely conducted in public institutions, is almost entirely organised in the laboratories of private corporations and driven by their particular interests. Furthermore, under the new rules governing international commerce, that research is considered to be confidential, a trade secret that is protected by laws and the regulations of the World Trade Organization and the Agreement on Trade-Related Aspects of Intellectual Property Rights (TRIPS) as well as the General Agreement on Tariffs in Services (GATS); those services are now deemed to include water, so that, for example, the results of testing bottled water are now also considered commercially sensitive – and confidential. On the same basis, seeds can now be patented and their use restricted, as Monsanto regularly and systematically does. If, as has been suggested, Monsanto is about to enter the field of water provision, it will control the entire agricultural cycle. But this time there are no pretensions to feed the hungry. The consequence of both Green Revolutions has been to drive small farmers off the land, to replace local farming methods – which took into account specific climate conditions, tended to grow a number of crops in their fields and adapted and responded to specific conditions – the new industrial agriculture is a monolithic, monoculture geared to global markets rather than local needs. Grain and rice are now dominated by global agricultural monopolies like Cargill. Tropical forests like the Amazon and the Indonesian rainforest are decimated to extend the land available for soya cultivation and cattle farming – Brazil is now the world's second largest soya producer, for example. It is also now the world's largest exporter of cattle and a major exporter of poultry, all of them intensively and industrially farmed. They are fed with alfalfa, a major product of California, and one of the world's thirstiest crops. Other Green Revolution crops include cotton and sugar cane, both of which absorb huge amounts of water compared with other pulses and food crops, while sugar and maize, as well as soya, are destined for bio-fuels. The production of biofuels, ostensibly to replace fossil fuels or at least a proportion of them, actually requires the massive use of fossil fuels and water. The reasons why their production was encouraged and accelerated under the presidency of George Bush is eloquent testimony to the fact that the central consideration was less the provision of alternative fuels than maintaining stocks of oil and controlling production. Given Bush's consistently hostile attitudes to the environmental movement and to the various earth summits, it is hard to reach any other conclusion.

The Earth Summit in 1992 returned to the issue, and advocated control and regulation of bioengineering and the protection of biodiversity. As we have seen, by this time, and in the wake of the Brundtland Report of 1987 on water, the depletion of water resources, their widespread contamination, and their loss worldwide could already be described as a global emergency. In the Punjab, for example, as Shiva reports, 10 per cent of the land was already saline and waterlogged, largely the result of dams, and 50 per cent damaged by a variety of other causes.[18] In the wake of the Earth Summit a number of governments, including India, acted to legislate for environmental controls and oversight. In the United States, however, the government acted quickly to deregulate – enacting into law a concept of 'substantial equivalence' which essentially removed any mechanisms for determining the impact of GM on any agricultural product. This gave the largely US-based agricultural and chemical multinationals carte blanche to drive their way into poorer economies, using the price mechanism to undermine local producers and the WTO to force open poorer economies to their cheaper products. Shiva gives the example of edible oils,[19] of which India uses a dazzling variety. In collusion with the Indian government, local oil mills were closed down overnight, supposedly on health grounds, and within ten years soya oil covered 70 per cent of the market.

If the test of the Green Revolution's promises is the disappearance of thirst and hunger from the world, then it has manifestly failed. But then, the purposes of the so-called revolution changed completely from its original expressed intentions. It was now a servant of a global trade order in which everything was a commodity – water, food, minerals, oil were all in the same order of things, their use and exploitation determined entirely by their price. Their human or social value, their capacity to answer needs or protect life was impossible to measure against market criteria. That is why the spread of GM crops could coincide with the declining calorie intake of Indian farmers and their families and why malnutrition can be on the increase. In Kenya's Rift Valley,[20] in Colombia and El Salvador flowers and vegetables are the major crops; 8 million hectares thus far have been removed from food production serving local needs to expand the cultivation of these commodities for sale in the West. Each of these commodities, and soya, wheat, cotton and meat, require large amounts of water for their production, water that is then carried in 'virtual' form to the water rich countries of the north. That is why there is still hunger and thirst, and as the land, exhausted and poisoned by the twin miracles of pesticides and GM seeds, becomes less and less productive and the water cycle increasingly disrupted, those problems are unlikely to be solved while it is global capital and its interests which prevail everywhere.

The carnivores

The most water-intensive of agricultural products is meat in general and beef in particular; it takes 15,000 litres of water to produce each kilo of beef (pork requires something like half of that). We will explore this when we come to virtual water, in Chapter 6. It is the case, however, that the production of meat has expanded exponentially with an enormous impact (40% of grain is used for animal feed and almost the whole global production of alfalfa.) The greediest in terms of input is beef. There is, therefore, a slightly edgy attitude on the part of many environmentalists towards carnivores, as if they were somehow responsible for the condition of the planet. And the same argument extends to diet in general. The argument is that individual food choices have a significant impact on agricultural production and in general on the use of water. This is true, of course, but set against the balance of use these personal decisions are important as part of raising the understanding of how the world works and as a way of encouraging people to support movements for change wherever they occur and on however limited a scale. We know now more than ever that a powerful living movement will connect the local and the global. The reality is that individual choices about diet will probably have less impact than the decision to form part of that global movement which can and will, eventually, change the world in a way that no individual action can do.

Figure 5.2   The disappearing forest. Sumatra. © F. B. Anggoro (Greenpeace)

In our view, and beyond the question of consciousness and political education, there are two ways of responding to the issue of diet, and particularly meat. It is alleged that the rising consumption of meat in the 'anchor economies' – that is, those en route to industrialisation – is the inescapable consequence of progress. But why should that be so? Protein has many sources, vitamins too – in other words, there are alternatives. Is there something inherently luxurious about red meat, or is it essentially the product of sustained, heavily financed marketing, which has successfully associated meat, and hamburgers in particular, with material progress? In a world where everything, including meat, was not a commodity, it might be possible to place before consumers a range of beautifully presented, tasty, fresh food to tempt the palate away from steaks. Meat consumption is rising in China, but for these same reasons. Why is the inverse assumption that a vegetarian diet as rich and varied and sumptuous as the Indian would simply be abandoned for a McDonald's? Is that not the arrogance of a Western world which identifies progress with its own specific way of life, even while it is critical of it?

Industrial water, industrial waste

Industry overall uses 25 per cent of the world's water; this is an average, of course, and the distribution of water between agriculture, industry and domestic uses varies by country, region and continent. In China it is 23 per cent, in India just 2 per cent, in Mexico 9 per cent, in Canada 69 per cent and in the United States 46 per cent.[21] In some countries, like the United States, Canada and most Western European countries, there are environmental regulations that constrain what industry can do with its waste, though not how much water is used as energy. Hydroelectricity, at least until the end of the era of dam building, was seen both as a cheaper and a less polluting alternative to coal and oil. We now know, and have discussed already, that that was extremely questionable, and that in various ways dams are responsible for greenhouse gas emissions as well as a range of other negative environmental impacts. In the developing world, such laws and regulations either do not exist or they are ignored or set aside. The palpable result is the huge proportion of industrial waste that is simply discharged into waterways and from there finds its way into underground aquifers and ultimately the ocean. The worst offenders in respect of CO_2 emissions are cement, oil, iron and steel, aluminium, bricks, pulp and paper and fertilisers.[22] Pulp for example uses 60,000–90,000 gallons of water per ton of paper; bleaching uses 48,000–72,000 per ton of cotton; a silicon wafer plant, incredibly, uses 4.5 million gallons per week.[23] In 1980, 237 billion tons of industrial effluent were released into waterways, in 2000 the figure was 468 billion tons. The sky-high levels of pollution in China's rivers are eloquent testimony to the absence of regulation.

There is one central fact about the industrial waste and its effects – that it increases with wealth. Thus while the overall figure is 59 per cent worldwide, 80 per cent is produced in the rich, developed countries, 22 per cent in the developing world and 8 per cent in the poorest countries. The projected figure for 2025 is that industry will be responsible for 24 per cent of the world's total freshwater withdrawal (that is, 1170 cubic kilometres[24] as compared to 752 cubic kilometres in 1995). The annual accumulation of heavy metals, solvents and toxic sludge is 300–500 tons. More than 80 per cent of hazardous waste is produced by the USA and the industrialised nations, where there are also environmental protection laws increasingly being passed to control that pollution. In developing countries by contrast 70 per cent of industrial waste is dumped into waterways and aquifers, contaminating the usable water supply. As with most of the other issues raised in this book, recognition of the problem has been slow to dawn on governments and international authorities. Of the 25 centres set up by UNESCO to monitor and research water related issues, 22 were set up after 2002. It would be optimistic to assume, however, that this will lead to a similar rush of action. A more immediate matter of concern is the fact that many major industrial companies are seeking ways to transfer their production to other, less developed areas, where environmental regulation is either lax or non-existent.

Two major industries responsible for a significant proportion of pollution are mining and oil.

Mining

While it operates on just 1 per cent of land, the mining industry represents 14.4 per cent of the world's economy, and in the last decade and a half it has become once again a major actor across the world. In part that is a response to the rising price of gold – currently at an astonishing $2,000 an ounce – perhaps as a result in turn of the falling value of oil reserves. In Peru, this has created a kind of gold rush, with major multinationals entering the boom but also an increase in small-scale wholly unregulated gold mining. Copper extraction is also at a peak, in particular to supply the construction boom worldwide, at least until 2008, and uninterruptedly in China. New areas of mining have arisen to feed the massive world mobile phone and electronics industries with the coltan and lithium that are essential to both. Eighty per cent of the world's coltan is in the Congo, where the metal is brought up from the ground by children in appalling conditions.[25] It is also mined in the Brazilian and Venezuelan Amazon though it is present in far smaller quantities. Lithium, which is used for the same purposes, tragically lies in large quantities under Bolivia's beautiful salt lake, the Salar de Uyuni, where Japanese, Russian, French and German companies are competing for access to it.[26] Diamonds too are mined in primitive conditions in several areas of the Amazon.

Every area of mining is destructive; it affects biodiversity chemically and sometimes by altering the physical environment, like the mountain-top removal method for mining coal which is practised in many areas of the USA. There is no restoration of the landscape when the coal seams are exhausted. All mining uses water in large quantities to process ore, and generate huge amounts of pollution, discharging effluent and leaving tailings (essentially slag heaps) which seep into the landscape and the waterways over a long period. The new drilling technologies now being developed are able to drill more low-grade ore, and leave behind them more waste and more contamination. The main forms of pollution are acid rock drainage, in which the contact between sulphides in the rock and water and oxygen produces sulphuric acid, heavy metal and chemical contamination and sedimentation as a result of land erosion. For example, for every tonne of copper removed from the ground, 99 tonnes of waste material are left behind; in the case of gold 3 tonnes of ore must be hacked from the landscape to garner the gold for a single wedding ring.[27] In Nevada, for example, the Humboldt River has been drained for gold mining – over 580 million gallons were pumped out between 1986 and 2001.[28]

In the current phase of aggressive mining, Canadian companies have played a particularly prominent and negative role.[29] Mining Watch reports

Figure 5.3 'Multinationals out', Quito, Ecuador. © Mike Gonzalez

the activities of Canadian mining companies in Africa and Latin America, as well as in Canada itself, where their record is extremely poor. It is not simply that the mines gouge out whole sections of the landscape with no thought of restoration; the companies systematically appropriate local water – by buying or leasing the land above it – and deal violently with protests by local communities. There are too many examples to list them all, but they include Excellon Resources in Mexico, Tahoe Resources in Guatemala, and Barrick – whose record is particularly poor – which operates in Papua New Guinea as well as Chile. Blackfire Exploration opened a mine in Chiapas, Mexico, ignoring local protests. They continued and the company closed its operations after the murder of a community leader. Canada itself has laws of corporate social responsibility, but they are largely observed in the breach in Canada, and in the external operations of Canadian mining companies, they are simply ignored. On the contrary, Mining Watch reports that the Canadian government's policy of what it euphemistically calls 'economic diplomacy' actually translates into active support, diplomatic and economic, for the companies. In Ecuador, in the area around Cuenca, Corriente Resources battled with local communities and organisations over its contamination of their water resources. One of the community leaders, José Tendetza, was killed in mysterious circumstances, when he was on his way to a major conference in Lima to expose the activities of the mining corporations.[30] The company then ceased its activities briefly, and was sold to what appear to be Chinese and Taiwanese interests, who resumed mining. We were able to see the aftermath of one Canadian company's activities in the Yasuni region of the Ecuadorean Amazon, which abandoned its mine leaving behind pools of contaminated water in local rivers and basins.

Peru, however, has the dubious distinction of having the highest concentration of large mining companies in any country – 14 belonging to the

International Council on Mining and Metals. The largest, the gold mine at Yanacocha in Cajamarca province, is jointly owned by an American company, Newmont Mining (the majority shareholder), Buenaventura (a Peruvian mining company) and the World Bank's International Finance Corporation, which has a 5 per cent holding. The history of the mine is one of continuous and often violent confrontation between the consortium and local communities.[31] In 2005, a mercury spill affected over 100 people, who later sued the company. A year earlier, the company's plan to extend the mine to the nearby Cerro Quilish brought sustained resistance as the mountain stood above the main water source for the area. Further protests held the project back until 2010 when it won government approval, but then further protests paralysed the project again. In 2012 a government environmental assessment gave the go-ahead but a local farmer, Máxima Acuña, has refused to sell her land, which stands in the way of the development, despite pressures and threats by the company and its private security force. Newmont goes to great lengths on its own website to represent itself as environmentally concerned; it claims to have suspended the project for the moment in order to build a dam and a reservoir. But these are not its concerns; its interests lie in the 6.5 million ounces of gold (worth at current prices, $13 billion) and 1,700 million pounds of copper within the mine. In Rio Blanco are mines owned by the Chinese company Zijin Metals, and the giant Rio Tinto Zinc, now Brazilian-owned and a renowned polluter, owns the La Granja copper mine in Lambayeque. In Andalusia, southern Spain, where the original Rio Tinto (Red River) is sited, the river is literally red because of the heavy metal and iron pollution that has seeped into the river from RTZ's mine workings.

Every mine project in Latin America has produced sustained protest and confrontation – which is why every company has its own security force. And in every case, be it in Peru, Ecuador, Colombia or Guatemala, water is at the heart of the struggle of local communities and organisations. And this is not mystical or resistance to modernisation, but a defence of the resources that sustain life and which are being removed to a global market without regard to any consequences, environmental, social or economic. It is worth remembering that 40 per cent of Peru's population and 70 per cent of its indigenous peoples live in poverty. Yet in 2007 mining produced $176 billion – 62 per cent of the country's total exports. It is obvious but worth restating that very little of the enormous profit that is and will be made from mining will accrue to the communities whose land it is, not even in wages and employment, given the notoriously poor working conditions in the mines. There is also a growing informal sector in and around the mines, whole families sorting through tailings or streams, the very streams that are polluted from mine effluent, searching for scraps of gold.

Big Oil

The other and probably more visible major polluter of the planet is oil. An industry dominated and controlled for most of its history by a cartel of oil giants – Chevron-Texaco, Exxon, Shell, BP. Despite the nationalisation of some major oil companies – Anglo-Iranian (Iran), Aramco (Saudi Arabia), PDVSA (Venezuela), Petrobras (Brazil), Petróleos Mexicanos and others – eight out of the twenty biggest companies in the world are oil and gas corporations. (Nine are banks). In fact the oil and gas companies control not simply the extraction of oil but also its refining, distribution and so on. And wherever world oil prices threaten to go up – as they did over several years up to 2014 – the largest producer, Saudi Arabia, will release more onto the market in the interests of big capital and big oil. China figures large in the Forbes list, with four of the nine banks and one of the oil giants (PetroChina). Their power and spread across the world has given them a prominent place in a different and much less flattering list – the world's major polluters. The most dramatic examples are a matter of public knowledge – the appalling sea bed explosion at a BP installation in the Gulf of Mexico in 2010, the Exxon Valdez incident in Prince William Sound off Alaska in 1989, the Ixtoc oil spill in 1979 of 140 million gallons of crude into the Gulf of Mexico, the Atlantic Empress that spilled 90 million gallons into the sea after colliding with another super tanker, the 1992 Fergana Valley oil spill in Uzbekistan, the Amoco Cadiz crash in the English Channel. There are many other spectacular incidents like these that hit the headlines, though most do not. But the real damage stems from the day-to-day, month-to-month pollution that is the inescapable consequence of oil and gas production. Between 1950 and 1972 oil production grew at a remarkably uniform rate of 9.55 per cent per year. This was not a happy coincidence, but a sign of the worldwide control exercised by Big Oil – the giant multinationals who control world oil production. Jonathan Neale makes a critically important point regarding Big Oil – that whatever they may say in their relentless public relations campaigns, they cannot go green. And that is not simply because the costs of conversion to wind or solar would be astronomical, but because their century-long monopoly is founded on an interconnected network of politicians, governments, banks and armies,[32] which are the guarantors of their continued power in the global economy. It is also the guarantee that while efforts have been and are being made to encourage them to accept environmental responsibility, even in the USA where several of them are based, major concessions have been made in the areas of pollution. In the developing world, there are few functioning controls or regulations. Wherever there is drilling, production leaks toxic crude into the soil and the groundwater, 'produced water' releases saline into the soil in large quantities, the familiar 'flaring', the flames rising into the air, produce the greenhouse gas methane, and air pollution is a major but rarely mentioned issue too (for oil as well as coal). In 2014, the USA announced that for the first time it was self-sufficient in oil (though that did not mean it would cease to be a major oil importer for the foreseeable future). What had made the difference was 'fracking' or hydraulic fracturing. Many governments enthusiastically adopted fracking, simply because it produced cheaper oil – or at least it did, if you considered only sales. The environmental impact, however, of smashing layers of rock with high-pressure water and chemicals, close to existing wells, changing the geology of whole regions and driving pollutants from the oil exploration directly into aquifers, is barely known, though there seems to be a consensus that it will almost certainly exacerbate seismic activity. The truth is that little is known in terms of its potential for polluting and creating major geological trauma. Scotland and the state of Pennsylvania in the USA announced moratoria on fracking.[33] Fracking will produce contaminated or saline water that will need to be disposed of elsewhere in deep wells; and there is a greater likelihood of earthquakes albeit of smaller magnitudes.

But normal oil production is heavily polluting. Take the example of the Niger Delta in Nigeria, 20,000 square kilometres of wetlands alive with agricultural products, wildlife, fish and a variety of plant life – or at least it was. Since the discovery of oil and gas deposits, largely in this area, toxic crude has leaked into the soil, the rivers and streams, and basins around the region, and the mangrove swamps, extraordinary instruments of ecological balance, have become contaminated. Like so many other wetlands on earth, some 50 per cent of them overall, pollution has devastated them and turned many of them into dustbowls.

And there is a further feature of the last 20 years, the period of accelerating exploitation and contamination of freshwater sources worldwide. In many countries, the state took over control of the oil industry. This was no guarantee of cleaner, greener production, nor of rational investment strategies – but it did at the very least make public accountability an issue, even if only in the denial. And however miserly the royalties (and they were rarely more than 10% at best) paid to states, they did add something to national budgets. Nationalisation of oil and mining monopolies was a major issue in many countries – in Libya (oil), in Chile (copper), in Venezuela (oil), among many examples. But the privatisation drive beginning in the late 1980s undermined any possibility of state control, as the international financial agencies pressed for privatisation. The example of Venezuela is interesting; there the renationalisation of Venezuelan oil in the 1990s was in fact a fraudulent exercise, leaving the financial and operational control with private companies while making the state responsible for the maintenance of infrastructure. The real nationalisation of oil, which was the central pillar of Hugo Chavez's government programme in 1998, finally happened in 2004, placing the state oil company PDVSA and all its revenues in the hands of the state. A contrary example is the Brazilian state oil company Petrobras. In 1995 legislation allowed private companies into the oil-producing sector, although only Brazilian companies (which were in reality Brazilian subsidiaries of the multinationals). The huge scandal that broke on the Brazilian public in 2014, involving corruption on a massive scale among its senior officials, often in collusion with banks, and also affecting the dam building programme, is an indication of what happens to the public interest in every area when the state sector is privatised.

Water flows through every sector of the globalised economy and its new areas of rapacious exploitation. Much of it is removed in different forms by the industrialised agriculture and industrial growth of recent decades; of the rest much is rendered undrinkable and unusable for the populations who remain, living beside the brightly coloured contaminated rivers or the piles of industrial detritus. Capitalism, in its furious race for profit and its competitive regime will leave destruction wherever it passes – if it is allowed to.

6

Virtual Water

Little attention has been paid to the fact that total water consumption and pollution relate to what and how countries consume and to the structure of the global economy that supplies the various goods and services.

The Water Footprint Assessment Manual[1]

A bad dream

Imagine a trip to the supermarket, anywhere in the developed world. You have a shopping list; the usual weekly items are on it. You will buy dairy products – eggs, butter, cheese, milk; meat, probably beef or pork, and almost certainly chicken in one form or another. There will be fruit and vegetables, some pre-packed and some loose – they may or may not carry their country of origin. Your potatoes are probably Chinese, your runner beans from Kenya, your rice from India or China or Thailand. You might buy flowers – from Kenya or Colombia most probably – coffee from Nicaragua, tea from India. And perhaps a new T-shirt while you're there. In the dream, you're alone; the long passageways are empty. You go to the fridge to get the orange juice for the kids and the semi-skimmed milk for their cereals. But when you pick up the carton it liquefies in your hand and what pours into your hand is not milk but water; the vegetables, the flowers, even the T-shirt turn to water in your hands. Perplexed and confused you go back to the car and drive to a local service station to fill your tank; but when you pull the trigger at the pump, all that comes out is water. You wake up with a start – damp with sweat.

Perhaps it was a nightmare, but the story it told was true. What you had imagined was a world in which all that you consumed was transformed into the 'virtual water' of which it was composed. You had seen, in vivid Technicolor, their 'water footprint' defined by Arjen Hoekstra *et al.*, who coined the concept, as 'the volume of freshwater used to produce a product, measured over the whole supply chain.'[2]

Table 6.1 Water Footprints

Product	Amount	Water equivalent	Global average	Notes
Apple	1 (150gms)	125 litres	822 litre/kg	
Apple juice	1 glass (200ml)	230 litres		
Banana	1 (200gms)	160 litres	790 litre/kg	
Beef	1kg	15,400 litres	15,400 litre/kg	Water footprint of its protein 6 times that of pulses; of calorie content 20 times that of cereals
Beer	1 litre	298 litres	298 litre/litre	Barley footprint 1,420 litres/kg
Bio-diesel (soya)	1 litre	11,400 litres	11,400 l/l	Soybean footprint 2,145 l/kg
Bio-ethanol (maize)	1 litre	2,854 litres	2,854 l/l	Maize footprint 1,222 l/kg
Bio-ethanol (sugar beet)	1 litre	1,188 litres	1,188 l/l	Beet footprint 132 l/kg
Bio-ethanol (sugar cane)	1 litre	2,107 litres	2,107 l/l	Sugar cane footprint 210 l/kg
Bread (wheat)	1 kg	1,608 litres	1827 l/l	The average is global. In Western Europe the water footprint of wheat is lower.
Butter	1kg	5,550 litres	5,550 l/l	940 litre/litre cows milk. Low compared with other animal products
Cabbage	1 kg	370 litres	270/l	Largest producers China and Japan. Footprint of Japanese cabbage is 130 l/l
Cheese	1kg	5,060 litres	5,060 l/l	940 l/1 footprint of cows milk

Product	Amount	Water equivalent	Global average	Notes
Chicken meat	1 kg	4,330 litres	4,330 l/l	11% of global meat production. Lowest water footprint of meat products
Chocolate	100 gram bar	1,700 litres	17,000 l/kg	Cocoa beans footprint 20,000 l/kg
Coffee	1 cup (125ml)	130 litres	18,900 l/kg	Roasted coffee. It used 3.7% of the total water flows in agriculture and industry together
Cotton	1 shirt / 1 pr jeans	2,500 litres / 8,000 litres	6,000–22,500 l/kg	The water footprint varies: China 6,000 l/kg, USA 8,100 l/kg, Uzbekistan 9,200 l/kg, Pakistan 9,600 l/kg, India 22,500 l/kg
Cucumber/pumpkin	1 kg	350 litres	350 l/kg	China 60% of total water footprint
Dates	1 kg	2,280 litres	2,280–3,650 l/kg	Dates from Iran and Saudi Arabia have a higher footprint
Eggs	1 (60gm)	200 litres	3,300 l/kg	Higher footprint for protein than pulses
Goat meat	1 kg	5,520 litres	5,520 l/kg	1% of footprint of animal production
Lettuce	1 kg	249 litres	240 l/kg	
Maize	1 kg	1220 litres	1,220 l/kg	USA 760 l/kg, China 1160 l/kg, Brazil 1750 l/kg, India 2,540 l/kg. 10% of total footprint of world crop production
Mango	1 kg	1,800 litres	1,800 l/kg	India largest producer, China/Thailand 1,430 /2,860 l/kg respectively
Milk	1 kg	1,020 litres	1,020 l/kg	Higher water content (31%) per gm of protein than pulses. 19% of total animal production
Olive oil	1 kg	14,400 litres	Olives 3,020 l/kg	96% for olive oil, Spain 2,750 l/kg, Tunisia 9,150 l/kg

Product	Amount	Water equivalent	Global average	Notes
Orange	1 (150gm)	80 litres	Oranges 1,020 l/kg	
Pasta (wheat)	1 kg	1,850 litres	1,827 l/kg	Varies according to origin of wheat
Peach	1 (150 gms)	140 litres	910 l/kg	Italy 420 l/kg, China (largest producer) 1,120 l/kg
Pizza Margherita	725 gms	1,260 litres	1,260 l/kg	Italy 940 l/kg, USA 1,200 l/kg, China 1,370 l/kg
Pork	1 kg	5,590 litres	5,590 l/kg	19% of total animal production protein water content higher than pulses
Potato	1 kg	290 litres	290 l/kg	China 22% of world potato footprint
Rice	1 kg	2,500 litres	1,670 l/kg in paddy	13% of world crop production; 22% blue water footprint
Sheep meat	1 kg	10,400 litres	10,400 l/kg	3% of total animal production
Sugar (beet)	1 kg refined	920 litres	132 l/kg raw	Average blue water content of cane 27%
Sugar (cane)	1 kg refined	1,780 litres	210 l/kg raw	
Tea	1 cup (250 ml)	30 litres	8,860 l/kg	India and China 51% of global footprint
Tomato	1 (250 gm)	50 litres	214 l/kg	
Tomato ketchup	1 kg	530 litres		
Tomato puree	1 kg	710 litres		
Wine	1 glass (125 ml)	110 litres	Grapes 610 l/kg Wine 870 litre/ litre	1 kilo grapes = 0.7 litres wine. France/Italy Spain water footprint 90/90/195 per glass respectively

Data from www.waterfootprint.org/ productgallery (accessed 17 March 2015).

The 'virtual water' or 'water footprint' concept is relatively new – it was developed in the early 1990s by J. A. 'Tony' Allan but only seriously addressed in water research during and after the International Expert Meeting on Virtual Water Trade held in Delft in December 2002. It is an evocative metaphor, and scientifically sound, but people should not be afraid that their favourite foods will suddenly melt in their hands. It is a way of showing where the world's freshwater (its 'blue water') and its rain and evaporation ('green water') has gone, how it has been used, and why we should beware of considering it an infinite resource. But that is not to say that we must abandon life as we know it, nor expect the earth to become a desert overnight. Hoekstra, *et al.*'s opening quote addresses *both* patterns of consumption *and* the global organisation of production and distribution, which is the area where the critical decisions about consumption, production and the distribution of goods (and wealth) are made by a very small proportion of the world's population. Together they make up the 'water footprint', a slightly more comprehensive concept than virtual water. This is an issue that needs to be properly and carefully discussed, walking the tightrope between an argument that we must all become vegetarians or abandon bathing on the one hand, and the recognition that world production is skewed towards some markets at the expense of others, and shaped by considerations of profit rather than the needs and desires of a humanity able to consider the options *without* the aid of advertising, product placement or the domination of public debate by interested – and very wealthy – parties.

Where the virtual water is

A good deal of it is inside us. To get a real sense of how water is used, or misused, it is useful to begin where we are. That is not to say for a moment that the solution to providing adequate water for all purposes for everyone on the planet will happen when we make small adjustments to our domestic usage. In the developed world, we typically use 300–400 litres each per day. Some we drink, by itself or with coffee, tea and so on. But that's a tiny quantity. Our 150 litres a day in the UK, or 350 in Australia, or more in many parts of the USA, is mostly used up in the toilet, the bath or shower, and in washing up. Beyond that, if we are fortunate enough to have gardens a lot goes there, a lot more if they are lawns. And then the really enthusiastic householder might use a good deal more washing the car – though probably that is largely to impress the neighbours. In fact, the supermarket dream should serve to bring home the fact that we use a great deal more, much of it taken from countries which may well have far less water than Europe, Canada or the USA.

The chart above gives some sense of what's involved. Today, for instance, a great deal of the water that is drunk comes in plastic bottles,[3] and that is not only in the developed world. In many third-world cities, public water supplies are unreliable and often unsafe, so there too bottled water is widely consumed, even though the price may be several hundred times the price of tap water. But the significant thing is that the bottles themselves consume 8 to 10 times their content in the manufacturing process. The bulk of our 'water footprint' comes in the food that we eat, however, as the chart shows. What does it actually represent? In the first instance, it is the water used to grow the crops themselves. In the case of meat and animal products, it is the feed (grain or alfalfa, for example) that is given to the animals, feed which has a far greater water footprint than pasture, which is largely watered by 'green water' (rain that is) rather than irrigation water from rivers or aquifers. But a large proportion of meat is currently factory farmed. This is reflected in the very high water footprint for all meat, particularly beef. Vegetables and fruit, pulses and some cereals are the least water-demanding of crops. The figures, of course, are an average, and in the real world they vary quite a lot according to the climate, soil and weather in the regions where they are grown. But the footprint also includes other elements – for example, the impact on 'blue' (fresh) and 'green' (rain) water, but also 'grey water', a calculation of the water affected by pollution, which is the amount of water it takes to dissolve contaminants. The important thing about grey water is that it is not returned to its source – it reduces the total of available freshwater.

Tony Allan, who coined the term 'virtual water', gives an assessment of the average consumption of people across the world,[4] which largely coincides with other writers on the subject. The average inhabitant of the developed world uses around 1,200 cubic metres per year. Allan helpfully suggests a visual equivalent – one cubic metre is about three full bathtubs.[5] The average US citizen uses around twice the average whereas in China the average falls to 700 cubic metres per annum. Our freshwater needs are around 200 litres per day – but if we calculate our virtual water needs, the figure leaps. Allan, for example, suggests a modest breakfast of eggs and bacon, toast, coffee (with milk) and an apple has a water footprint of around 1,100 litres. The largest water footprint in Western culture, however, is beef (at around 15,400 litres per kilo); other meat has a significant footprint, but none more than one-third of beef's impact. There is a common argument that the consumption of beef is a major cause of water scarcity in the world, particularly because India and China – the largest of the developing world's industrialising nations – are turning to meat consumption on a large scale. Vandana Shiva has no truck with that assertion as far as India is concerned; she points out that Indian cuisine is a vegetarian cuisine, that it is a profound cultural tradition rather than (as seems to be the implication) an unfulfilled yearning for a Big Mac! After all, even India's wealthiest families are vegetarians. There is evidence of an increase in beef consumption in China, and we know that the arrival of the hamburger chains in Japan, which barely knew what heart disease was because of its almost entirely fish and vegetable based diet, has unfortunately introduced a range of meat-related health problems into the culture. It is a complex question, of course, but there is no *automatic* or *necessary* connection between rising standards of living and beef consumption; it is a created need, a product of a highly organised consumer culture that associates burgers and steaks with prosperity.[6]

Pearce estimates the global water trade to represent about '1000 cubic kilometres a year or twenty River Niles'. Of that two-thirds is in crops of every kind, a quarter in meat and dairy products and a tenth in industrial products. The agricultural production also includes bio-diesel and bio-ethanol, which add their enormous production water footprint to the total trade. The movement of virtual water, of course, is in both directions. The biggest net exporter of virtual water is the USA, much of that in grains either directly for consumption or as cattle feed (100 cubic kilometres annually). A great deal of that is grown in the High Plains and watered by the huge Ogallala aquifer, which is alarmingly depleted. Brazil is now the world's largest exporter of beef, largely from an Amazon region whose rainforest has been torn down to make way for the expanding soya bean plantations and cattle ranches.[7] Who imports the virtual water? Major importers include Japan and the European Union, who have no water problem at all; China, in addition, is importing increasing amounts of grain directly and in the form of meat. But much of the food consumed in the Middle East, where water is scarce, is imported.[8]

In the opposite direction, the West imports its virtual water in the form of food – rice, fruit, coffee and tea, sugar – and cotton. Cotton is a particularly vivid example of how the virtual water trade works. Grown in India and China, the raw cotton is exported to Malaysia, for example, a major centre for the manufacture of cotton textiles, from where it is sent on to T-shirt buying markets of Europe and the USA, Egypt and Sudan, Pakistan and the Central Asian republics. The Aral Sea and the Indus River, which no longer reaches the sea, are casualties of the cotton trade. The Rift Valley in Kenya no longer feeds its population but is a centre for growing flowers for export. In terms of a rational and ecologically sound use of water and land, the virtual water trade is a massive paradox. The oranges exported from southern Spain and Israel,[9] the strawberries of which Mexico is the world's largest exporter, the tomatoes from an Ethiopia regularly affected by drought, are all products of dry areas sent to water-rich countries and exacerbating their own water crisis. Why not grow crops for local consumption rather than grow them for export to the opposite side of the world?

It is a very obvious question and the answer is equally plain – the profits earned by a global export trade dominated by a small number of multinational companies. The decision by Brazil to resume the deforestation of the Amazon Basin after a five-year moratorium, for example, has nothing to do with the internal needs of the population of the country but with its ambition to become a major power in the global market. The mining and oil companies that are drilling their way down into the watershed or into the mountain sources of Amazon water export their profits to their home base while leaving behind a minimum gain for local states. This systematic global exploitation, which often goes by the name of neo-liberalism, gave rise in its turn to a reaction among the countries of Latin America which had suffered most from the international strategies of the World Bank and the International Monetary Fund – Venezuela, Bolivia, Ecuador, for example, led by Venezuela in the early years of the twenty-first century.

The neat pack of chilli peppers or green beans on a supermarket shelf will now probably carry an indication of their country of origin and possibly of their calorific content; they may well bear the seal of organisations that confirm that organic farming methods were used, as the Soil Association does in Britain. But it will as yet carry no indication of its water footprint.

But even if and when it does, it will need some explanation of what it means. Cotton and rice continue to arrive from India and Pakistan, from fields irrigated from rivers and increasingly from underground fresh water sources. As Fred Pearce points out, the feared famines of the 1960s did not happen on the scale that were anticipated; today the world produces twice the amount of food that it did then but with three times the amount of water. It is an obscure statistic – perhaps as difficult to grasp as the population figures or the expanse of water in cubic kilometres. But it is far easier to understand when presented with a picture of a dried out river, or the declining water level of lakes and waterways.

Water, as we have said several times, is mobile; it moves from place to place, changes shape and speed of flow. But this does not explain water poverty and water wealth in the contemporary world. There are countries, desert nations for example, with very limited water supplies. Yet ironically enough, these very countries export their own water in virtual form, in many cases as oil. Yemen and Saudi Arabia are two of them.[10] Tony Allan explores the case of Spain in great detail, for although much of the country is arid, it has exported and imported its water in such a way as to maintain its water use at a consistent level, restraining its own water use and importing virtual water from elsewhere. It can do this, as Allan points out, because it is a modern and diversified economy (despite its post-2008 crash problems). The Middle East, North Africa and much of sub-Saharan Africa, however, export their water in a consistently unequal exchange; so too do most countries in Asia and Latin America. In Europe and North America, by contrast, there are increasing regulations governing water use – which are largely, and unfairly, compensated by increasing virtual water imports. Someone once said that 'water flows uphill, towards money'; presumably it was virtual water they had in mind.

In India, the ground water revolution since the 1980s has seen at least 21 million wells drilled, with a million more added each year, and a doubling of land irrigated by them [...] About 250 km^3 is extracted every year, 40% more than is replaced by rain [...]

Farmers in Saudi Arabia used up more than half of a groundwater reserve of close to 500 km^3 to produce wheat at subsidized prices [...] in Mexico 80% of all water is used by agriculture and ground water contributes 40% of it. Irrigation supports more than half of all farm production and three quarters of exports, most of which […] are for the U.S. market.[11]

These examples can be repeated over and over again. The pattern they expose is of a poor world forced to export its water in forms determined by the global market and its main actors. Monsanto in pesticides, Cargill in agricultural land, the five giants in oil, the members of the International Mining Convention, the Chinese banking sector, Wal-Mart and General Electric. And the prices they are paid for their products are determined far from their point of production. Until 2000, the price of coffee was largely determined in London, at the London International Financial Futures and Options Exchange (LIFFE), which employed the method called 'outcry' in which people in multi-coloured blazers moved around the floor shouting. To the observer it was incomprehensible, but it was a very long way from the coffee growers around the world who saw only a tiny proportion of the money exchanged between the voluble traders, whose earnings would have been equally beyond their understanding. The poor, by contrast, can move only as transient and immigrant labour, the most demeaning of all human conditions, restricted and limited by immigration authorities, exploited by people traffickers, and set to work at starvation wages. It is a situation whose ethical emptiness Tony Allan justifiably compares to the slave trade. It is interesting to note that calculation of the water footprint does not include labour. Industrial capital and its financial backers, by contrast, can and do move their production facilities effortlessly across frontiers, in search of the subsidies often paid by the state in developing nations to attract their investment – and those will usually include water, energy and infrastructure. The water companies themselves illustrate the point well, since they are all very reluctant to take responsibility for infrastructure, preferring to leave that to municipalities and governments while reaping profits from water distribution. So water becomes one more commodity whose value is its price, the returns to be gained from it, rather than its contribution to human life.

The developed world imports its virtual water in order to allow capital to invest in more lucrative uses for its water – industrial or manufacturing – or in water supplies to urban centres. The international agencies impose methods and technologies that are produced in the industrialised economies – like electrical water pumps for tubewells – as well as the production priorities for agriculture. The result is not the alleviation of hunger in the developing world but the creation of deeper and wider divisions within them. It is rare for the press not to carry reports of Chinese and Indian multimillionaires and their extraordinary lifestyles. It is far less interested in the fact that in 1991 Indians ate 177 kilos worth of calories per day, whereas in 2007 the level had fallen to 152.[12] California exports ten cubic kilometres of virtual water per year; yet it is an arid state, whose consumers use double the global average of water.[13]

Where does the discussion about virtual water take us?

Trading virtual water

In the recent public discussion, virtual water has been part of a campaign directed fundamentally at individual or family consumers in the developed world as an ethical question. And for some people, it has led to a personal decision to take measures to limit water consumption – turning off taps, installing water-saving sanitation devices, deciding on purchases on the basis of their virtual water content, or becoming a vegetarian. As an expression of environmental consciousness that is legitimate. But it would be naive to imagine that decisions of that kind, multiplied even by millions, could do more than make a tiny impact on water distribution on the planet, even within the 10 per cent of world water consumption that could be affected. For Tony Allan, who coined the concept, its significance was slightly different. He saw the calculation of virtual water as a solution to a worldwide social problem, and indeed as a way of preventing the water wars which by the end of the twentieth century were becoming more and more a topic of discussion in the public arena.[14] He argued that nations should export products in which they have a comparative advantage and import those in which they have a comparative disadvantage,[15] which in his view would produce real water savings.

This seemed to be an argument against subsidising agriculture in arid countries, like the MENA countries of the Middle East. Implicitly it was also a critique of an idea, which has been fundamental in discussions in the developing world, namely food self-sufficiency, or food sovereignty, as a primary objective. A report commissioned by the Australian state of Victoria, however, offered a very critical view of the idea.[16] In the first instance, water is not in that sense portable; what is saved in the exchange cannot then be delivered to the beneficiary of that saving. And that is true both in the international exchange and in the exchanges that might occur within countries or regions. Water not used in one activity may not be usable in a different activity; and the non-water elements involved in the production in the importing country – transport, for example, employment patterns, not to mention cultural issues – may very well simply not be transferable to a different activity. As the report's writers put it, 'the virtual water concept offers no effective guidance to policy makers regarding either water use efficiency or the sustainability of water use.'[17]

The average figures in our chart are therefore probably not very helpful, in that the same product produced under different conditions may produce a very different water footprint – say beef produced in Botswana and Holland. And there are major critical social and cultural issues which may well be very hard to quantify, but which are centrally important. For example, in the case of the arid countries of the Middle East, a dependence on water imports, actual or virtual, will be very likely to carry a political implication, perhaps of dependence on a single product – oil, for example – and by extension on a single market. There may also be a significant effect in social terms; where the crop which is replaced by its imported equivalent effectively undermines a local agricultural economy, with all its human and cultural consequences. The problem with the discussion of virtual water trade is that these issues (which Hoekstra, *et al.* do allude to in the quote that opens this chapter) are rarely addressed in discussions of a virtual water trade, which seems to be largely focused on quantitative criteria and market considerations.

A major study conducted in Germany addresses these issues in admirable detail, and we have relied on many of its findings in what follows.[18] Virtual water trade usually occurs between the North and the South – in other words between developed and developing or less developed nations, where the exporters are subsidised by home governments. The USA, for example, is the largest exporter of grain to the world, much of it from the High Plains area, the bread basket of America, where producers enjoy significant government subsidies. On this basis alone, this is not an equal exchange of like for like. It is true that many developing countries will also subsidise their agricultural producers, but the scale and conditions of production will inevitably be very different and to the disadvantage of the poorer countries. Furthermore, the subsidies will have very different purposes, as the operation of the World Trade Organization has made abundantly clear, in using its considerable economic weight to undermine subsidies or protective tariffs in non-developed areas to the advantage of the major, northern-based players.[19] Virtual water exports may make sense for a group of countries in process of industrialisation (the authors call them 'anchor countries') like China and India, to the extent that it will enable those countries to shift finance from agriculture to industry, where the yield on virtual water will be significantly greater – that would seem to reinforce Allan's view of the beneficial effects of virtual water trading. Yet even here there are very important caveats. It is unlikely in general that water can simply be transferred from one sector to another, at least on any reasonable time scale. An example might be China's South–North Water Transfer Project, taking virtual water from the north to the south and then returning water to the industries in the north. The Belo Monte Dam in Brazil, transferring water from the watershed to new industries in northern Brazil, could be another. But in both cases, the time-scale is long and the costs, financial and social are, as we have discussed, enormous. Given the impact of industrial pollution on the one hand, and the general underdevelopment of wastewater treatment on the other, the argument that virtual water trading is a conservation measure falls beneath its own weight. In any event, global conservation is an entirely abstract concept since water saved in one place cannot, by and large, be delivered to another.[20]

The calculation of the value of virtual water is itself a vexed area, since water has no cost as such to be included. One way of assessing its market value might be to calculate what the cost of producing a given crop might have been in the importing country as opposed to the exporting, but that is problematic too. The cost of cultivation is not the only factor; there is also a question of how that (blue and green) water might have been re-used after cultivation, and the social costs which we have already referred to. The German study argues that an essential consideration in determining (non-economic costs and benefits) must be the ability of the importing country to absorb the farmers no longer working in agriculture into other employment sectors. If that is not addressed then the new impact in social terms of the trade will be to create a new class of workless economic migrants.

In less developed countries, where foreign exchange is scarce and the infrastructure is likely to be poor, the virtual water trade makes even less sense. The authors suggest, with supporting evidence, that this kind of trade is conducted between major global economic actors on the one hand, and governments on the other. The result is likely to be that the distribution of the imports will be largely limited to the cities – bringing no benefit to the rural populations who no longer produce the food crops – and determined in a centralised way that isolates the farming population even further and undermines any form of decentralised control of production or indeed of water.

There may be some justification for short-term virtual water trading in response to some specific set of circumstances, but that could produce the problems that food aid has exposed, undermining local production and reducing the prices that farmers can charge for their products in the face of large-scale, uncosted imports (though we know that in one form or another aid is always paid for by the recipient).

There is an argument to be made for virtual water trading *within* countries or regions, as in the Southern African Development Community (SADC) or the Latin American regional organisations like Mercosur or ALBA, but only to the extent that local conditions, traditions, cultures and needs can be included in the trade relationship. We are not here discussing virtual water trading in a vacuum, but in the context of the globalisation of water. As we have seen in the cases of mining and oil, as well as cash crops for export, globalisation does not permit trade as a system for resolving needs – as is suggested in what Boelens, *et al.*[21] refer to as 'neo-liberal utopianism'. It exists to sustain and deepen an unequal relationship in which the process of aggressive accumulation, the Holy Scripture of capitalism as Marx put it,[22] redistributes wealth nationally, regionally and above all globally to the global corporations.

WATER BEYOND THE STATE: A LETTER FROM COCHABAMBA*
Marcela Olivera

Coordinating Committee for Water and Life, Cochabamba

The traditional forms of organisation in Bolivia are characteristically autonomous and horizontal. Together they constitute a real, practical and day-to-day means of understanding the public good and of living out a participatory democracy beyond the limits of the state and the government of the day.

---

* A longer version of this article was published in *NACLA Report on the Americas.*

▶

The water committees of the south district of Cochabamba, Bolivia's fourth largest city, are the epitome of these horizontal and autonomous organisations. This still very active network came to public notice after the 'water war' of 2000, when a mass popular mobilisation stopped the attempt to privatise the municipal water system.

While many people associate the water war with the idea of an authentic democracy, this could be contradictory; a war – any war – implies violence, the loss of energy and resources, discord and death, while the function of democracy (as we understand it in the West) is exactly the opposite – to avoid them. But the battle was not simply a defence of this resource. One of the main impulses behind these struggles was the historic and permanent struggle of Bolivian men and women to defend their right to decide in a horizontal and autonomous manner what their needs were, in other words, the urgent and enduring need to live in a true democracy.

In September 1999, the Cochabamba Municipal Drinking Water and Sewage Company (SEMAPA) was sold to the Aguas del Tunari consortium, whose majority shareholder was the Bechtel Corporation. It was the latest in a long series of structural adjustments conceived and backed by the World Bank and the International Monetary Fund and imposed from the 1980s onwards on the countries of Latin America, including Bolivia. After the privatisation the residents of the Cochabamba valley saw their water charges rise at the same time as the cooperatives and the water committees had to administer water services with *no* state support, which meant that Aguas del Tunari could take them to court for unfair competition and seize their installations. This was the grim situation that was faced by the Coordinating Committee for the Defence of Water and Life (known as the Water Coordinating Committee – or *La Coordinadora del Agua*). After several months of negotiations with the state and confrontations with the armed forces, the Coordinadora was successful in driving out the company.

Many aspects of our reality became visible for the first time as a result of the Cochabamba Water War of 2000. For example, a multitude of forms of organisation that were not run according to the structures of Western democracy, one example of which were the Cochabamba Water Committees, which were central to the Coordinadora. After the confrontations in 2000, they joined with other sectors who actively participated in those events and established coordinating networks.

The Cochabamba water committees are traditional in the south of the city, but they also exist in the peripheral areas. The southern district consists of six separate areas that cover half the city's population – more than 200,000 people. The district has between 100 and 120 water committees, in addition to the 400 which exist across the whole of the

▶

metropolitan area of Cochabamba, according to Stefano Arcidiacono of the NGO CeVI. Together they represent thousands of people organised, though not exclusively, around water issues.

There are no two services in Bolivia that operate in the same way, but what is clear is that organisations like the Committees share a vision of water as a living, divine being, and as the foundation of reciprocity and complementarity. It is a being that belongs to everyone and to no one, an expression of the flexibility and adaptability that helps nature to create and transform life and ensures social reproduction. The assemblies of the Water Committees reflect the traditional values and usages (*usos y costumbres*) of the community, reproducing in the city communities like those that exist in the countryside.

Historically, water in Cochabamba has been distributed in a number of ways. Although the municipal company was the main provider for the urban area, almost two-thirds of Cochabamba's communities developed their own systems based on rivers, wells and rainwater harvesting. Some communities raised money to build a water distribution system and contributed regularly to its maintenance. Other districts can have access to them through water tanks or even through private water tanker lorries. The decision about how to gain access to water is one of the things that has made possible the building of the commons.

Many people have attributed the origins of the autonomous practices adopted by the water committees to the Inca Empire, via the colonial period to the present day. The result is that the water committees are often seen as contemporary expressions of ancient communal practices. In his 2001 article 'State and autonomy in Bolivia, an anarchist interpretation', Carlos Crespo explains that in Bolivia autonomy is 'not an ideal to be pursued but a day-to-day practice of people, communities and affinity groups.' These horizontal processes have always been the social and political practice in Bolivia in relation to the State and the governing power. This was reflected in a declaration published by the factory workers during the water war, under the title 'Neither public nor private but self-organization'.

Like the indigenous struggles and the organisation of their societies, the water committees represent an anti-state vision of autonomy, since they have emerged from the marginal zones around the city, the so-called poverty belt, where there coexist peasant immigrants, who have brought to their new urban communities the traditional Andean practices of shared labour by rotation, known as *ayni*, and relocated miners who brought to these communities abandoned by the State all the experience of union organisation in the mines.

▶

The water committees are the result of the promotion and strengthening of the people's autonomous forms of organisation. They are based on practices that are not recognised by the state or the international community – but nor do they need to be. The community members divide up and share roles in responding to the question of how to provide water for their neighbours, building networks that allow them to organise together and share strategies for providing water to their communities. They are not organised *against* public water systems but *in favour of* developing the capacity to decide how and how far they should connect with those public systems. They are clearly, then, the most authentic expression of how to operate politically in an autonomous way.

For example, at a meeting in 2008, the members of an area in the south of the city decided to divide up their research into how best to deliver water to their district. They then shared their findings and discussed whether it would be best to pressurise the local municipal water company to lay pipes to their area or to build a giant water tank and negotiate with the water tankers to keep the tank filled on a fixed contract.

In the water committees the most important social concerns have to be addressed by the community as a whole and that is what distinguishes them from the state's perception of social movements as expressions of demands made on the state by the community; in the water committees *people organise to decide and create the conditions of their own life* and not to beg favours from the governments of the day. That is why they go beyond the question of the absence, shortage or abundance of water to address a range of other issues: the welfare of community members, security, festivals, football matches and so on. As one committee member, Gastón Zeballos, explains: 'If someone dies, for example, we discuss making a donation to the family, or at New Year we give a basket of basic necessities to every committee member. We even discuss personal issues in our assemblies.' Such is the degree of independence of action of the committees that they have been called 'areas liberated from the state'.

The current situation in Bolivia is confused and the challenges faced by the committees are many, from their own technical and financial limitations to their negotiations with the state over their autonomy. The technical issues stretch across the formation and consolidation of the water committees in Cochabamba; people are aware of their limitations, as the contribution of Zeballos, leader of the San Miguel Km 4 Committee, explained to the First International Exchange of Experiences between Water Operators from Uruguay, Colombia and Bolivia (URCOLBOL) meeting in Montevideo in October 2013. During the discussions at the meeting, the Bolivian participants were more interested in technical questions of chlorination and

▶

the treatment of drinking water and wastewater than anything else. As Zeballos put it, 'we are more interested in technical issues because we've got the social indicators covered. In our committees there is participation, social control and rotation of duties. It's the other stuff we need.'

Economic limitations are as important as technical ones, especially in relation to absolutely necessary projects that cannot be completed with local resources alone, as is the case with sewers in the area covered by each committee which apart from cost have to be seen as part of a global system covering both municipal and alternative systems. This type of project requires state support and investment, a support that should include the willingness to respect the autonomy of the water committees, their particular vision of their own needs; as well as contributing to methods of managing and distributing water that are accessible to all, at election times they can become clientilistic.

When Evo Morales took power in 2006, there was a hope that his government would extend the autonomy and self-organisation of the social movements. In fact, the opposite has happened. The state has expanded into new areas, one of which is water, and there is an increasing tendency for the state to intervene in areas that were traditionally outside its control. The result is a high level of centralisation. Recent water legislation has given the state power to take decisions on and intervene in community systems and autonomous practices that it did not previously recognise.

Through the Convention on Water Rights promoted by the Bolivian government in the United Nations and the Universal Declaration of the Rights of Mother Earth, the Morales government declared that nature has rights and created a definition of 'rights' that passed the responsibility and therefore the power to administer water from the people to the state; these decisions have been celebrated internationally, and Morales is considered a leader in environmental reform. The traditional ways of determining the use of water have thus been declared null and void, and those who seek access to water must now turn to the state, the law and the courts.

Throughout their history, the water committees have faced technical and economic challenges and continuous attempts at incorporation by the State. But their success, epitomised in the water war, has shown that by organising horizontally Bolivians can recover the right to administer their common goods autonomously, against the established power and the traditional ways of understanding democracy. Today the people are not organising to make demands of the state, but to shape and determine the conditions of their own life.

7
Water and Global Warming

All climate havoc touches our lives through water and snow.

Vandana Shiva (2011)

In 1989 a large group of scientists from all over the world set up the Inter-governmental Panel on Climate Change (IPCC); it also included politicians from sympathetic governments. As Jonathan Neale points out, they were neither environmentalists nor radicals, but scientists 'concerned to alert the world to what was happening'.[1] Emissions of greenhouse gases, principally carbon dioxide and methane, were raising the earth's temperature, with consequences that could be imagined if not fully predicted, but all of which would have serious negative effects on the planet and its inhabitants. The main producers of these greenhouse gases were car exhausts, industrial production, particularly oil, gas and coal, as well as methane produced in agriculture and in water, particularly as a result of evaporation at dam heads. The US delegation, however, included representatives from oil and coal companies like Exxon and Peabody who mounted a counter-campaign, arguing that global warming was not happening, and that any attempt to limit CO_2 emission would adversely affect American industry. This was hardly surprising given that the USA was responsible for over a quarter of global emissions, as well as being a dominant influence in the UN, the World Bank and later, most significantly, the World Trade Organization.

There was no doubt, however, that there was a highly organised and well financed campaign to discredit the climate change theory; the science was sound and the 1990s brought more and more evidence to the table, evidence that was sufficient to convince a number of European governments to pull back from attacking the IPCC, but which failed to counter the relentless campaign mounted by American business, working through organisa-tions like the George C. Marshall Institute[2] and other well funded friends of industry. Fred Pearce gives a useful summary of what global warming is.

Certain gases in the atmosphere, most importantly water vapour and carbon dioxide, trap heat. They have a 'greenhouse effect'. This in itself is no bad thing. Without those gases the planet would freeze. There is also no doubt that human activity – mostly burning carbon-based fuels like coal and oil, but also deforestation and ploughing – is pumping carbon dioxide into the atmosphere [...] We would expect these gases to warm the atmosphere, and the atmosphere in turn to warm the oceans, starting at the surface [...] On its own, a doubling of carbon dioxide levels in the atmosphere would raise global temperatures by an average of only 1 degree C [...] But the evidence is that this will be amplified [...] by 'feedbacks' [...] the most obvious of them [is] the melting of ice [...] Ice is a superb reflector of solar energy; it bounces right back into space. Less ice means more solar radiation is absorbed by the dark surface of the oceans and land and radiated back into space as infrared heat.[3]

The process in its turn will produce a build-up of water vapour, which will add to the heating, though by how much is still a matter of argument. But there seems to be a consensus that on current trends the planet's temperatures will rise by between 2 and 4 degrees Celsius this century. There is no doubt that they will rise; the issue is by how much. The Climategate affair gave a new boost to climate sceptics, at least temporarily; but the political manipulations behind their claims, and the shallowness of them, are also more transparent now than they were, as the social movements of the late twentieth and early twenty-first century have exposed the role of organisations like the WTO and made very public the machinations that have limited the outcomes of the climate conferences. In Copenhagen in 2009, for example, the United Nations Climate Change Conference ended in disarray with a weak statement on global warming drafted by the USA, China, India, Brazil and South Africa which was merely 'noted' by the conference, but not adopted. The conference was also marked by the exclusion of critical voices from Latin America, namely Hugo Chavez and Evo Morales. According to documents published in 2014 by Edward Snowden,[4] it emerged that the US and Chinese delegations were party to information obtained by spying on other conference delegations!

The issue is no longer whether there is global warming, but what measures governments are willing to take, separately or together, to address the problem. What the Copenhagen Conference revealed was the degree of resistance on the part of big capital through the US and Chinese delegations. They were completely unwilling to contemplate limiting emissions.

Figure 7.1   Coal mining is polluting China's Yellow River.
© Zhu Jic (Greenpeace)

Pearce is right to insist that while the effects of global warming now confirm many of the predictions of the climate researchers, it is not always possible to predict where further changes will come, nor exactly what is responsible for them when they come. But we do know that a number of rivers have seen a reduced flow – like the Indus and the Colorado, the Rio Grande in the USA or the Murray in Australia. Lakes fed by these rivers have in their turn shrunk or dried out. As the air warms, evaporation will increase and the amount of water vapour in the atmosphere will also rise – Pearce estimates up to 8–10 per cent more, or as he puts it 'enough to fill twenty Niles'.[5] The consequent increase in rainfall globally will change climate in many places. One effect will be an increase in the ferocity and speed of cyclones, which will deposit their rain load in different areas. Whole regions will grow drier and others will experience unpredictable and unexpected rainfall. It is projected that the reservoir levels on the Colorado River will fall, as will the great rivers that bring water to some of the world's poorest areas. Like the Niger, they will lose a proportion of their flow.[6] The fate of the Amazon and the Orinoco is still unclear, but the northern rivers of Canada and Russia will swell as the air warms and deposits more rain in them. The melting glaciers around the world will, in the short term, add to the flow of some rivers. But as they shrink, so too will the rivers they feed. The cities of La Paz (Bolivia), Quito (Ecuador) and Lima (Peru) will each face a water crisis as the melting continues. And 85 per cent of the Himalayan glaciers, whose snow melt feeds the rivers that sustain over half the world's population in India, China, Pakistan, Bangladesh and the whole of South East Asia,[7] are in retreat with consequences still difficult to predict.[8]

The direct effect of global warming is already clearly perceptible in the Arctic regions. The thickness of the ice was reduced by four feet, and by 2007 the loss of undersea ice was also increasing with the result that major ice formations like the Greenland and Ward Hunt Ice Shelf are starting to crack and fragment. The permafrost, the snow cover, in the Arctic is also slowly melting, releasing methane as it does so. According to Alan Anderson, this process of thinning is continuing and will continue.[9] As it does the potential wealth of the Arctic will become accessible for the first time. Once again, and in keeping with the culture of neo-liberalism, what is a disaster for mankind in many ways (like water scarcity) is seen, within the narrow focus of a global system based entirely on profit, as a commercial opportunity. Its result is a race to control these new potential resources and the markets they will command.

Five nations have coastlines bordering on the Arctic regions – Russia, Norway, Denmark, Canada and the United States. As the ice melts, and the rich store of minerals beneath the ice cap – oil, gas, diamonds, platinum, nickel and lithium –become accessible, the remorseless battle to exploit the situation for profit accelerates. There is oil under the Beaufort Sea, and the United States has already reserved eight plots beneath the ice. The Arctic is governed by the UN Convention on the Law of the Sea (UNCLOS), which established that each nation with an Arctic coastline shall have 370 kilometres of territorial waters; what remained would be international waters to which no country could lay individual claim. After ratification of the Convention, however, each country would have the right for the following ten years to claim an extension – for example, Canada and Russia claim that geological exploration has shown that their continental shelf extends beyond the 370 kilometre limit. The object of the exercise is neither conservation nor the development of policies to try to detain, or at worst slow, the process of global warming. But then you would hardly expect that to be the concern of the companies who are now seeking to exploit its enormous potential wealth – companies like the oil giants Shell and BP, India's huge iron and steel multinational ArcelorMittal or Australia's Greenland Minerals, whose eye is fixed on the platinum and nickel deposits. The north-west passage, the emblematic channel for commercial mariners for several centuries, has now (according to a 2007 European Space Agency report) thawed to the point of allowing free passage; competition between shipping companies vying for the transport franchises has intensified as a result. What could be seen, and should be seen as a warning sign to the planet in an age of globalisation is interpreted only as a commercial opportunity.

Anderson ends his book with his own warning of what we might expect if things continue this way. The Russian town of Norilsk,[10] within the Arctic Circle, was once inhabited only by Soviet prisoners; its current population of around 230,000 live in one of the world's most polluted cities, and one of only two (the other is Yakutsk), in a region of permafrost. Temperatures there reach 50 degrees below zero and it has 45 days of total darkness annually. There is not a single tree within a 48 kilometre radius, the consequence of the acid rain resulting from the massive quantities of sulphur dioxide (2 million tons annually) and copper and nickel oxide (500 tons) pumped into the atmosphere during the processing of the nickel which is its principal industry. Its inhabitants have a life expectancy ten years below the Russian average (at 46). We may take this as a nightmare image of an Arctic future under global capitalism.

The Copenhagen conference mentioned earlier ended in confusion and with an Accord which was binding on no one. Thus the world's principal polluters and the world's most powerful economic actors conspired to make pious statements while accepting no controls whatsoever on their activities. Their governments, after all, saw themselves as the advocates of their country's private business interests, as we have seen, for example, in the case of the Canadian state's vigorous defence of its major mining companies. Multinational capital drives the World Trade Organization and the World Bank and the international financial institutions like the World Bank and the IMF reciprocate by tying all their loans to states and governments to privatisation – as was the case in the imposition of tubewells in India, of private corporate involvement in water distribution across the world from South Africa to the Philippines, of the building of dams across the world's major rivers.

Earlier in the decade, the central focus of climate change debates was carbon dioxide emissions into the atmosphere. The solution to the problem offered by governments and corporations was carbon trading; an economic solution to a major social and environmental problem, as usual. In fact, of course, all it meant was that polluters were required to pay for their contamination of the environment, or they could buy carbon credits from others who did not have enough capacity to pollute themselves. The paradox was that the largest recipients of pollution *rights* were those most responsible for it – so the Indian steel giant Mittal, with something close to a monopoly on global steel production, was able to buy 1 billion pounds-worth. And we now know that the extension of carbon credits for hydroelectrical projects is fuelling the new wave of dam construction in Brazil, financed by China whose industry will benefit from the production of iron and aluminium in the Amazon.[11] Vandana Shiva's comparison with the purchasing of indulgences from the medieval church to pay for sins committed seems peculiarly appropriate.

Today it is only the most extreme of climate change deniers who would question the fact that global warming has occurred because of greenhouse gas emissions into the atmosphere. We have addressed where these emissions come from in several places through this book. The main industrial polluters are the energy industries – fossil fuels above all, gas and coal, as well as the cement, wood pulp and steel industries, and of course auto production. But although most people would not immediately associate agriculture with greenhouse gases, we now know that pesticides and herbicides and deforestation are major generators of greenhouse gas emissions. In her 2011 Oregon World Affairs Council Lecture,[12] Vandana Shiva calculated that 14 per cent were directly produced by industrial agriculture, 18 per cent by deforestation, 7 per cent in the transport of food and 3 per cent in manufacturing food packaging, most of which was simply waste.[13] Still less would water appear on a common sense list of sources of carbon and methane; yet as we discussed earlier, dams in particular generate both methane and carbon dioxide as their plant life rots or de-oxygenates in the accumulated silt at the dam heads.

For all the public discussion of global warming, there seems to be a reluctance among large numbers of people to see it as a current problem. It is as if global warming will come upon us at some future tipping point, when the earth's temperature hits a rise of 3 or 4 degrees. In fact global warming is a process of change, a series of partial steps towards the more catastrophic events anticipated in some of the more apocalyptic climate change warnings. But global warming does not start then, but now; it is manifested in separate events across the world, whose connection may not be immediately obvious. What, for example, are the links between forest fires in Russia in 2010 and the six-year drought after 2000 in the Murray–Darling area of Australia? How do floods in the Ganges Basin and drought in the Amazon connect? What has the Indian super-cyclone of 1998, whose velocity doubled that of earlier cyclones (from 150 to 300 kilometres per hour) to do with Katrina in 2005? The answer in a simple sense, is that they were all unusual weather events; each in their own environment indicated serious changes in weather patterns. They were in some ways unpredictable, on the basis of previous patterns; but the possibility of their occurrence, if not the likelihood, was in most cases predicted by climate scientists and meteorologists as the appropriate conditions were beginning to show. The Russian forest fires happened in an area which would normally have been cold at that time of year; the drought in Australia affected an area of high agricultural production, and almost destroyed production there; in India it is usually the Indus that floods rather than the Ganges, and the Amazon drought of 2005 and 2006 was unprecedented. Cyclones of increasing destructive power have been a growing feature of recent years across the world. Yet each was not simply a natural disaster, but a crisis produced by human intervention in nature – the rise in the average temperature of the earth arising from greenhouse gas emissions, the accelerated deforestation of the Amazon which led to the droughts of 2005 and 2006 as water vapour clouds sped over the rainforest and deposited rain on the Andes. And Katrina, presented to the world as a tragic accident, was the result of bureaucratic indifference and inadequate construction to save money, which was then diverted to the war in Iraq. As Neale reminds us, the US Army Corps of Engineers were highly respected worldwide for their skill and knowledge; it was they who built the original dykes (levees) that were supposed to protect New Orleans. But as federal budgets were cut, the Corps were replaced by subcontractors without the knowledge or experience, who built levees with weak or inappropriate materials, presumably while charging for the best. And an added risk came from the oil and gas exploration in the Mississippi Delta which had eroded the wetlands there that had acted to hold back previous hurricanes. The conditions were ripe for what local scientists called 'the Big One', an especially destructive hurricane that they and the Federal Emergency Management Agency (FEMA) had already predicted in 2001. But nothing was done, the warnings were ignored, and it was the poor of New Orleans who paid the price. That particular scenario played out within the USA, but it was repeated across the world.

Global warming, then, means climate change and instability; weather events that are difficult to predict but that will change the conditions under which crops are grown. And as Vandana Shiva noted, those events will express themselves through water or snow.

The reality, however, is that 'climate havoc' does not touch everyone with the same ferocity. It is a cliché to point out that hurricanes and cyclones cause far less damage, and far fewer deaths, in the USA or Europe than in the developing countries. And this is not, by and large, because these extreme climate events are in themselves more powerful or destructive in India or Pakistan or the Sahel than on the Florida coast. Katrina was a rare demonstration of that. But what happened in New Orleans exposed failures of a different kind – bureaucratic, administrative and above all a failure to provide adequate resources to address a tragedy that *was* foretold. The issue then is to understand why the poorer regions suffer more. In the Indus and Ganges Basin, for example, the canals and dykes were constructed to channel water towards areas of industrial agriculture; but the consequences were not considered, with the result that flooding and drought are occurring in these areas with greater frequency and with more damaging results. It is the *combination* of climate change and other cumulative issues which presents the most serious danger, a danger to life but also a serious risk in shortfalls of food production, as patterns of rainfall or lack of it change quite suddenly, affecting harvests. The rising temperature in the northern United States, for example, will affect the winter wheat crop; the drought resistant crops now being produced by Monsanto and others may, as has proved to be the case, fail to grow in normal conditions. The economic arrangements that enable multinational corporations to dump subsidised agricultural products on poor countries, bankrupting local farmers, but also undermining their domestic markets – either through WTO pressure or simply because local growers cannot compete with the price of imports – will expose those markets to catastrophic shortages should climate change affect the exporters, because of the depletion of the 4 trillion tons of water in the Ogallala aquifer for example. In the USA, 200,000 wells are pumping out the water at a rate of 50 million litres per minute – 14 times its recharge rate. The exploration of tar sands for oil and fracking, which will mix oil and water, in search of new sources of fossil fuels will cause damage we cannot yet predict.

A comprehensive review of the impact of climate change on China[14] suggest the best policy might be 'hope for the best, adapt for the worst' – which is not very reassuring. What the study shows is that recent weather events have indicated shifting climate patterns – heatwave events across China, decreased rainfall in the south-east and north-west, for example, and an 80 per cent depletion of glaciers which will reduce the supply of fresh water over time and more immediately may lead to lake outbursts and flooding. In agriculture, the authors point to increased production of some key crops, like rice in the north-east, but also that 'pests and diseases may also expand their geographic ranges as the climate warms, increasing stress on crops.' There is clearly a great deal of research under way in China, including the largest biotechnology plant outside the USA; but the solutions the authors suggest – increase in fertiliser use, more intensive irrigation and the introduction of new crop varieties seem fraught with their own dangers against the background of what the authors acknowledge is an unstable and unpredictable climate change scenario.

The Slovakian hydrologist Michal Kravcik has been instrumental in defining the problem, and driving home its urgency.[15] He describes the transformations of rivers as they are straightened and their meandering corrected, accelerating run-off with the result that the land absorbs less. In this way large amounts of water are lost to the ecosystem, leading in its turn to the drying out of river banks and a lower level of evaporation – in the long term, because without water the temperature of the land rises, this will produce desertification. He has mapped these places on the earth's surface where water has disappeared; he calls them 'hot stains'. His warnings of 'global collapse' may seem apocalyptic – and they are certainly futuristic and possibly the claim is excessive. But the processes he describes are manifestly real, as Juan Pablo Orrego discusses.[16] In the USA for example only 2 per cent of rivers are still free-flowing; the rest are dammed, and as we now know, dams release both methane and CO_2 from rotting vegetation and rising silt levels. For Kravcik, and others, irrigation is a central problem; China, the USA, India and Pakistan have over half the world's irrigated land and all of them are encountering problems of desertification (because of the salinity of reservoir waters), erosion, drought and water shortage. Twenty per cent of China's irrigation water comes from aquifers, and from the 2 million wells pumping water into the fields. The problem is not this or that specific difficulty but that these phenomena reflect an interruption, or rather disruption of the water cycle, short and long. As temperatures rise, evaporation occurs more rapidly and there is less water to fill the natural reservoir that is the soil itself. The deforestation of river banks and the extensive damage caused by cattle raising in the same areas contributes in its turn to the loss of run-off and the warming of rivers.

Something like 50 per cent of the world's population now live in cities. In the industrial world cities grew gradually over time, and water provision – at least from the mid-nineteenth century onwards – was integrated into that growth. Today's third world megacities grew over relatively brief periods, and in an explosive rather than a gradual way. Water provision was not seen as central to the ideology of twentieth-century urban growth, and today those cities contain large populations living in essentially fragile, improvised accommodation – or in the worst cases in cemeteries and on rubbish dumps. For them water is available only in bottles or plastic bags from water sellers who charge dozens if not hundreds of times the cost of tap water. There is a clear class dimension; the huge bottled water trade provides for the middle class, many of whom have water in their homes but very few of whom would drink it. The Millennium Development Goals, of course, focused on drinking water, but there is an ambiguity in the insistence on 'access' to water and even more so in the concept of 'improved sources' – improved from what might be the question to ask! Access might include standpipes or intermittent supplies; what is not specifically addressed is the specific duty of society to supply drinking water directly to people in their homes. In the same way, while international campaigns, including the Millennium Development Goals, have defined adequate sanitation as a toilet in the home, the problem of disposal remains unaddressed; in 90 per cent of these cities, human waste flows untreated into streams, rivers, lakes and aquifers, contaminating the available potable water supply. It may be, therefore, that other forms of disposal are more appropriate in the many places where wastewater disposal is a problem. It is an illustration of a principle that we have repeated throughout this book – that it is the people who are directly affected who know best, and it is their experience and preferences that should drive the solutions. The United Nations, in its water reports regularly reaffirms the desirability of 'participation' in projects. But that needs to be more clearly defined so that it is communities themselves who participate in decision-making rather than their often self-appointed representatives. And in addition to that, leaks (and theft) of water account in many cities for between 40–50 per cent of the available supply. It is a loss that hides behind euphemisms – like 'non-available water' – but is water of which the human urban population is deprived. The combination of contaminated dam water and untreated waste not only itself generates enormous levels of greenhouse gas emissions, it also reduces the water that is available to the human population.

Modern cities, wherever they are in the developing world, model themselves on the high-rise, concrete and glass utopias of modern urbanism. Streets and buildings are constructed using the ubiquitous (and greenhouse gas producing) concrete. The rain that falls on these great con-urbations has nowhere to go but into drains and sewers where it joins the flood of sewage on its way to the outlying sumps and sinks from where it will find its eventual way into rivers and aquifers. If it were allowed to penetrate the soil, it would intensify the water cycle and contribute to its regeneration. In Los Angeles, an emblematic representative of the concrete city, some streets have been restored to stone or gravel which are permeable, and this thirstiest of cities can return some of its huge consumption to the earth. Berlin, too, has changed the constitution of its streets to allow water to return to the water-table beneath. Paradoxically in the megacities, home to politicians and parliaments and corporations, money will be found to resolve at least some of the problems. But in the new generation of cities, urban centres with 200,000–500,000 inhabitants, it may be more of a battle to ensure provision of clean, safe water.[17]

Yet in all of these examples, the problem is not scarcity, but waste and mis-management. If the leakages were dealt with, there would be ample water for all urban citizens.[18] And Tokyo, Phnom Penh, Zaragoza and Singapore have shown, with their extremely low rates of leakage and rapid response to them, that a properly managed water distribution network can function well. The essential problem is a failure to manage the system professionally, honestly and with knowledge.

It is also perfectly possible to reduce the use of water in industrial and agricultural processes; Professor Asit Biswas reports[19] that US steel production has reduced by 90 per cent the amount of water it uses as compared with 1975 – although that will in part be because a great deal of steel production has moved away from the USA and Europe to the Far East, to China, and to India (the world's largest steel company, after all, is Indian owned). He asserts in the same lecture that urban water provision in the USA is still inefficient and could be reduced by a further 30 per cent. The point, as we have argued throughout this book, is that water depletion and pollution are aspects of the same process that has produced global warming. To conserve water on the one hand, and to redistribute it more equitably on the other, demands that both these processes have to be radically addressed, and stopped. This may, of course, sound utopian. But as we will go on to argue, it can be done. The resistance to controlling pollution and the misuse of the planet's water resources does not come from those suffering the consequences of both, as the increasingly positive figures for environmental regulation from the USA and Europe demonstrate, but from the corporations and global interests who are the beneficiaries of both – the agro-exporting industry, the manufacturers of pesticides and herbicides, the mining corporations, the oil and gas corporations, the car manufacturers and the banks and financial institutions who support them.

> Unfortunately, however, in many developing countries, the quality of hydro-meteorological networks has deteriorated, rather than improved. Unless this trend is reversed, there is a real danger that many developing countries may be stampeded into using many theoretical, untested and questionable methods, developed by the academics of the developed world, which are likely to be of very limited relevance or use for the conditions of the developing world because of the very different social, economic, climatic and institutional conditions.[20]

Climate change is transforming those conditions, rapidly and in unexpected directions; and they are affected by, and the result of, other factors (or 'externalities' as they are sometimes called) – population growth, technological advances, future food requirements, the future role of oil and gas and of alternative energy sources, and economic growth. None of these factors can be predicted with any precision. Yet each, individually and in relation to one another, will shape that future. If the 'academics of the developed world' are not in a position to know, recognise or respond to those changes, who is?

Faced with the uncertainty, which is perhaps the most significant effect of climate change and global warming in specific relation to water, Claudia Pahl-Wostl returns to a concept first coined by Professor Gleick in 2003. He referred to the 'soft path' in water management 'that complements centralized physical infrastructure with lower cost community-scale systems, decentralized and open decision-making, water markets and equitable pricing, application of efficient technology, and environmental protection.'[21]

Put in a different way Claudia Pahl-Wostl refers to a process of 'learning to manage by managing to learn', a cooperative and decentralising method which bears very little relation to the current realities of the machinery of world trade or the global balance of power in an age of globalisation but instead recommends a move from technical management to an integration of the human dimension. In short, a more adaptive and flexible method under fast-changing conditions, and under the control of its users.

8

Ya Basta! Enough is Enough!

Chiapas to Cochabamba … and beyond

Two sparks relit the fires of protest across the world in the 1990s. One was Cochabamba, the city in eastern Bolivia whose name (it means the lake on the plain) would become synonymous with the struggle for water as the twenty-first century began. The second, the Zapatista uprising in southern Mexico in 1994, is not associated in the public mind with the question of water. Yet for a new generation it exposed the realities of globalisation and the neo-liberal order. When the anti-capitalist movement marched against the World Trade Organization in Seattle in late 1999, and exposed its existence to a world that hardly knew of it before then, many of the demonstrators wore the red bandannas that had come to symbolise the Zapatista movement, and carried their slogan on hand-painted banners – 'Ya basta', 'Enough is Enough'. The connection between the two movements was not only water – though it was significant that while Chiapas supplied a large proportion of Mexico's hydroelectric power, half of its population had no access to fresh water. What they shared was a common enemy, and a common purpose – to resist a predatory neo-liberalism whose driving impulses were summed up by David Harvey in the phrase 'accumulation by dispossession'. The Cochabamba Water War,[1] as it came to be called, placed the issue of water on the agenda of a rising movement of anti-globalisation activists. For the first time, global capitalism and its instrument of expansion – the World Trade Organization – were planning to seize the commons, the resources which were a common property, or so it was assumed until then.

Water was not the specific issue in Mexico around which the Zapatista National Liberation Army (EZLN)[2] raised its banners of resistance on 1 January 1994. It was the communal right to land, to the *territorio* – a term that means more than simply territory, since it embraces culture, tradition and the collective *relationship* with the land. The historic form of communal ownership of land in Mexico was called the *ejido*; its rights were enshrined in the Mexican constitution of 1917 which protected that common ownership. Neo-liberalism, however, saw those historic rights simply as constraints on the free operation of the market in land and in the water rights that came with it. The formation of NAFTA, the North American Free Trade Agreement, required that those rights be removed and the land and water of the rural communities become available to the market. But the Zapatistas were also fighting a more distant but extremely hostile enemy. The Zapatista communities grew maize (the staple of the Mexican diet) on small plots which survived because they were subsidised through various state programmes and because tariff barriers ensured that the local product could compete on the internal market with the imported maize coming from the world's largest maize producer, the United States. The date of the Zapatista rebellion was well chosen; it coincided with the press conference at which the presidents of Mexico, the United States and Canada were to launch NAFTA, which was intended as the first of a series of regional trade areas. The world's press had been summoned to the capital, Mexico City, to hear the joint declaration of the three presidents and to applaud the onward march of globalisation. Instead the world's front pages pictured the poorest community in Mexico raising its weapons in protest and exposing the real effects of neo-liberalism. For the removal of tariff barriers and an end to subsidies were conditions imposed by the World Trade Organization, which would of course destroy the fragile livelihood of the peasant farmers of Chiapas. Water was not discussed specifically at the time, but looking back it is clear that this was an early chapter in the history of virtual water trading which we have discussed in Chapter 5. That is one connection. The other is that both heralded popular risings against globalisation after nearly a decade of programmes of 'structural adjustment' whose effect was to impoverish even further the developing world to the benefit of the interests of capital.[3]

But it was Cochabamba that created a global awareness that water itself was under threat of privatisation. From 1990 onwards, the drive to privatisation across Latin America was destroying the fragile public sector. In Argentina the auction of state assets under the hammer of Peronist President Carlos Menem sold off all the country's public owned assets to private corporations at bargain prices. Telefónica of Spain took over the telephone networks, the national oil company YPF went to Spain's Repsol,

Figure 8.1   Cochabamba: the 'water war' that drove out Bechtel and reclaimed water for the public sector. © Tom Kruse (Cochabamba Coordinating Committee for Water and Life, Cochabamba, Bolivia 2000)

the aeroplane construction company to Lockheed Martin and so on. And in 1993, the Buenos Aires water company was sold to the French-based multinational Suez.

In Bolivia the privatisation process was conducted by the military government of Hugo Banzer.[4] The privatisation of water was imposed by the World Bank as a condition for providing loans for the Bolivian economy; the argument offered, ironically enough, was that Bolivian state organisations were inefficient and prone to corruption.[5] A presidential privatisation decree in 1985 covered everything including water. In Cochabamba, Bolivia's third city, the municipal water company was sold in 1999 to Bechtel and the company Aguas del Tunari in which it was the majority shareholder, with a guarantee of a 16 per cent rate of profit. Bechtel was one of the six engineering companies that built the Hoover Dam in the USA in the 1930s, and it was also contracted for the construction of a dam near the city. Their name would become much more familiar worldwide in 2003 as one of the beneficiaries of the destruction of Iraq.

Bechtel immediately imposed major price increases on the city's population, even claiming that rainwater would now have to be paid for under a law which gave them exclusive rights to distribute *all* water.[6] Jim Schulz provided direct evidence from city dwellers that the new price structure meant that a worker on the minimum wage would be required to pay 20 per cent of his wages for water. At the very end of 1999, protests blocked the highways into the city; they brought together the poor communities of the city, the workers in the mainly small factories around the city, students, and the coca farmers (*cocaleros*) of the region around the city. Many of them were originally inhabitants of Andean communities who had been given small grants of land to grow coca (a legal crop) after the mining industries (principally tin) on which they depended collapsed in the 1970s. But they brought with them long histories of resolute and militant struggle, as well as a fiercely defended cultural and social identity, which gave those struggles cohesion and roots. The Coordinating Committee for Life and Water (*La Coordinadora*) was then formed; its secretary was an unassuming worker in a shoe factory, Oscar Olivera. Protests and occupations of the city square brought together a range of popular, grass-roots organisations. In February 2000, troops sent to crush the protests opened fire on the demonstrators. But the protests continued in a virtually continuous occupation of the city. In April, the government withdrew Bechtel's contract and control of the water company was passed on to a committee of local people. In Oscar Olivera's words, 'it wasn't just a victory over one company or even privatization as a practice [...] it was a process of emancipation for us all.'[7]

The subsequent history of the new municipal water company, SEMAPA, is sadly not as positive as we might hope. The creation of a Ministry of Water, under Abel Mamani, in 2006 was a hopeful sign, and the affirmation of water as a right and one that could not be conceded to private interests was a vindication of the demands that came out of Cochabamba. But there was a further step to be taken; the central state had to enter into a direct and living relationship with the grass-roots organisations – what Bolivian vice-president Álvaro García Linera has called 'a creative tension'. Indigenous organisations in particular have evolved complex systems of water distribution which connect with their cultural traditions and history; they have also evolved considerable skills in dealing with centralised organisations whose models of organisation derive in the majority of cases from concepts of development which are large-scale, industrial and consonant with the global organisation of trade. The tension here is between global formulae and the realities of local conditions, and it is at the heart of the argument about water. Local knowledge in this context is not simply about folklore

Figure 8.2   Man over nature; the destruction of mountains by coal companies.
© Wayne Pace (Greenpeace)

and traditions, but a resource that has enabled and will in the future enable
adaptations to shifting climatic and soil conditions.[8]

The global market, through its international instruments like the World
Trade Organization, sets out to impose universal solutions which clash with
local initiatives and experience. It does so remorselessly, using the concept
of free trade to open access routes into every economy and to overwhelm
them with the weight of its resources. This now extends beyond actual
product into intellectual property rights, which have allowed Monsanto,
for example, as well as others, to patent seed varieties and force farmers to
abandon the use of their own developed seeds; the attempt to patent spices,
for example, led to a long battle between the WTO and the government
of India over the ownership of turmeric, though in the end the Indian
government won the battle – having paid the expensive lawyers on both
sides who managed the case. Meanwhile the government of California
has sued British Columbia over its refusal to sell its water. Under WTO
rules, corporations who are refused access to natural resources can sue a
national government for anticipated profits. That is the case of OceanaGold,
a mining company refused a concession on the grounds that the mineworks
would pollute the drinking water of El Salvador. The dispute will now go
to ICSID (International Centre for Settlement of Investment Disputes), the

internal tribunal that resolves international disputes of this kind. This was the body to which Bechtel turned after losing the Cochabamba contract. The proceedings of ICSID are secret, and are taken by three judges; since the court has no policy on conflicts of interests, the members of the tribunal are almost all corporate lawyers.[9]

In an interesting article on water struggles in Latin America, José Esteban Castro[10] distinguishes two kinds of protest movement around water. One, which he calls 'ecocentric', revolves around the protection of ecosystems, local wildlife, the protection of biodiversity. These are generally directed against dam projects, and have little to say on the subject of the privatisation of water. The second group, the 'anthropocentric', focus on social issues of access to water, the accountability of water providers, pollution and so on, with less emphasis on dams and other large scale projects. The 'ecocentric' movements, however, for example in India and Pakistan, and Brazil, are more than simply environmentalist – they also invariably have a cultural component, a defence of both land and water as cultural values, as captured in Latin America by the difference between 'land' and 'territory' (*territorio*). And the anthropocentric will focus, in the case of major projects, on the huge scale of displacement that they will often imply. Yet the division between the two is often blurred, especially as recent water movements have learned from each other and assimilated one another's experience. In Brazil, for example, the anti-dam movement has embraced both the demands of the indigenous peoples of the Xingu threatened by Belo Monte, and the broader question of the accelerating destruction of the Amazon in the interests of industry and export agriculture.

The Indian constitution explicitly states that 'no citizen shall be subjected to any restriction with regard to the use of wells, tanks [and] bathing ghats [i.e. riverside washing areas].' Although the courts have reaffirmed the right to water as an aspect of the right to life, however, there is no legal obligation on government to provide water for the citizen. As elsewhere, globalisation and neo-liberalism began to take root in India from 1990 onwards, with policies that laid particular emphasis on water privatisation. New Delhi, for example, has failed to privatise its water supply, despite several attempts to do so, largely because of mass protest movements. According to Professor Asit Biswas, 99.9 per cent of the waste of New Delhi (the original city now lies largely underwater), some 200 million litres per day are discharged, untreated, into the river Yamuna;[11] the river, as a result, has been declared biologically dead, even though it is clean and clear just a few miles before it enters the city. The city uses more water per day than Canada, yet access is limited to three hours per day. The majority of its 18 million residents, however, have to queue for water sold from tankers, water that is notoriously smelly and contaminated. Although it is illegal to pump groundwater from aquifers, it is the norm in the city, and is used by both private and government bottled water plants. Yet 52 per cent of its piped water is lost in leaks and theft.[12]

In Mumbai, in 2006, under pressure from the World Bank, a pilot study was conducted with a view to privatising the city's water supply. The lack of transparency of the whole operation, and the absence of any democratic input, raised concerns in the city and a protest movement successfully stopped the attempt at privatisation by the back door. A new proposal by the City Council, however, argued in favour of putting in water meters in the slum areas; the social movement that had built up around water rejected the proposal, since it could only lead to cutting off people who could not pay – in direct contravention of the constitution's recognition of the right to water.

In Peru and Ecuador, the struggle over water is not simply about water; it is also about land and more extensively about democracy and rights. Ecuador as part of the progressive Bolivarian Alliance (Alba) includes in its 2008 Constitution clauses on water rights as well as several articles acknowledging the equality of all ethnic and community groups within the country. The concept of the good life (*sumac karsay*), that is a life lived in harmony with the environment, is also enshrined in the document, which also lays down that there should be *auditorías de agua* in every community – local committees elected to assess water use and regulate it. Yet the Water Law of that year did not exclude privatisation of water. Hence the fact that water in Ecuador's principal port and industrial city, Guayaquil, is privately owned and run by a subsidiary of Suez, Interagua. And in fact the state has signed concessions that include 2,240 million cubic metres of water – 74 per cent to the electricity sector, 20 per cent in irrigation and 1.22 per cent for domestic users.[13] The costs of water, however, fall disproportionately on the poor. Big companies pay very little; the peasant farmers of Guayas pay 120 times more for their water than the banana company Reybanpac or the San Carlos sugar complex. In Cotopaxi they pay over 50 times more than landowners and in Chimborazo 35 times more. The biggest producers, 1 per cent of the productive units, have access to 67 per cent of the water, while the peasant farmers, who are 86 per cent of users, can use just 22 per cent of the irrigated land and 13 per cent of the water.[14] The principal users are export agriculture, and increasingly mining. Food production is declining and the expansion of mining, and the pollution it entails, is depleting the freshwater stocks.

The resistance to this inequitable distribution of water and to the increase in mining activity, as well as oil concessions, has been led by the two main indigenous organisations in Ecuador, CONAIE and ECUARUNARI. The Shuar nation in particular has been active in the Ecuadorian Amazon and the Andes where their *territorio* is located. In Peru, the often extremely violent confrontations between communities and foreign mining and oil companies have intensified and multiplied in response to the expanding presence of multinational corporations across the country which we discussed in Chapter 5. Their struggles have found echoes in Colombia, and particularly in the Amazon region of that country, where flower growing for export and the use of river water by Coca-Cola and others have polluted the rivers and undermined the water supply for local farmers. Here too the confrontations have been violent with a rising register of deaths and injuries.

The plurinational constitutions of Bolivia, Ecuador and Venezuela acknowledged the citizenship of the indigenous people of those countries, not only as individuals but also as communities whose territories were guaranteed as a collective right. Yet in the second decade of the twenty-first century, which gave its name to this new current of political thought (twenty-first-century socialism was the term coined by Hugo Chavez of Venezuela), in Bolivia and Ecuador indigenous communities are in confrontation with their governments precisely over the issue of collective rights. It was one thing to recognise, and indeed to celebrate in these new constitutions, the cultural diversity that existed within these nation-states. Their declared multiculturalism expanded the cultural horizons of the nation and embraced the artistic and cultural expressions of each community, including the notion of 'territorio' as a category embracing culture, history, traditions, but also social practices and forms of organisation and the framework of law and custom in which these communities functioned.

However,

Although the neoliberal project speaks of decentralisation and respect for multiculturalism, the latter is allowed so long as it does not get in the former's way. Respect for certain cultural traits such as old-fashioned valuing of 'folk' culture is no problem; but if the idea is to manage resources differently than market logic, that is another matter.[15]

The fundamental problem for the new governments was that 'territorio' was a physical category as well as a cultural one – it was *land and water*. Furthermore, it was land, as we have seen, in the regions most lusted after by the multinational corporations seeking the new commodities – the minerals, the oil and gas, and the water – with which the territories were best endowed; they were in the Andean region and in Amazonia. Quite clearly, there was a clash between what rapidly emerged as the economic strategy of these 'new left' governments and the rights embedded in their constitutions. Despite their hostility to neo-liberalism their economic programmes continued to be based on extractive industries producing for the global market, albeit that they were negotiating for a more legitimate share of royalties from them to fund the social reforms to which they were also committed. Rhetoric and practice rapidly diverged.[16] The case of the TIPNIS national park clearly exposed this contradiction;[17] and although the projected highway was suspended, it was relaunched in late 2014 and will go ahead, despite the resistance it has met. And the Bolivian government has now allowed the Brazilian oil company Petrobras to drill in Tarija, near the dry Chaco region, and a Chinese oil company, Eastern Petrogas, to drill in the Aguaragüe National Park. The sophistry that land and water are distinct is meaningless of course, though it is one that well suits those who are attempting to ride an impossible paradox.

Government and companies consistently seek to separate the two. They might do this through natural-resource specific legislation that treats the resource separately from the territory, or through efforts simply to undermine territory. In either case, the effect is the same – to produce water and land as alienable commodities rather than as parts of territory.[18]

This goes far beyond issues of respect for cultural rights; it touches the very core of democracy. In Ecuador two-thirds of the Oriente region and three-quarters of the Amazon region have already been subdivided into exploration blocks allocated to multinational mining and oil companies, most of the latter in high mountain regions which contain the headwaters of the rivers of the Amazon. That is the basis of the struggles of the indigenous organisations across the whole region, in Ecuador, in Colombia, in Peru and throughout the Amazon. It is a bitter irony that those who recognised their rights in the abstract are attacking them when they practise them.

In South Africa, the struggle over water has formed part of a broader movement around the role of the state in providing services, especially to the poor and the working class.[19] A leading force in these struggles has been a social movement called the Anti-Privatisation Forum (APF), which has led protests and actions around not just water but also electricity provision and housing issues. The long struggle against South Africa's apartheid regime will be familiar to many people outside the country because of the success of the Anti-Apartheid Movement and the boycott campaigns, and because of the stature of Nelson Mandela as a symbol of that struggle. Despite the cruelty of the apartheid repressive apparatus, the black resistance movement was remarkable for its militancy and its consistency, as well as for the creativity of its protests. When the leading organisation in the struggle, the African National Congress, finally came to power in 1994, it formed a government with support of the South African Communist Party and the Congress of South African Trade Unions (COSATU) and has governed continuously since then. It presented its aims and intentions in a document called the Reconstruction and Development Programme (RDP), in which it undertook to meet the basic needs of South Africa's people and in particular to provide 'lifeline supplies of water and electricity to the poor', that high proportion of the population that was unemployed or working in occasional, precarious employment and which the white racist governments had permanently denied the facilities and resources with which to meet their needs. The RDP enshrined a concept of development which, in Marcelle Dawson's words, meant 'an investment in social capital', the improvement and enrichment of the life of the majority through the provision of services and access to education. Its promise, as it was understood, was that those whom apartheid had excluded from society would now have the opportunity to become full and active citizens of the new South Africa.

Two years later the ANC produced a very different plan, GEAR (Growth, Employment and Redistribution), which focused on privatisation and on the priority of 'full cost recovery'.

For many ex-ANC activists the document was a bitter disappointment that reneged on the progressive promises of the RDP. It abandoned in reality if not in words the redistributive horizons of the RDP, replacing them with criteria of financial viability. In a South Africa still deeply divided and profoundly unequal, it was a charter for business and a route map for a new black middle class, but for the poor it promised nothing. In Johannesburg, the immediate consequence was a municipal water company (JOWCO) run jointly by Sondeo, a subsidiary of Suez and two subsidiaries of RWE/Thames Water. The Anti-Privatisation Forum (APF) brought together over 20 organisations and others in response to the new proposals; its central argument was that water was a right, a public good and not a commodity that had to be paid for. The APF had its roots in the anti-apartheid struggle, and would use much of its repertoire of methods of resistance. In fact, the APF differed from other social movements at the end of the twentieth century in its political orientation. Its hostility to commodification and the domination of market criteria came from a clear political perspective. When JOWCO began to introduce pre-paid meters for the poor, the APF bitterly opposed them. This was not electricity, for which there were alternatives; water was irreplaceable, and the meters penalised the poor. It was true that there was some emergency provision, but it entailed attending the City Council offices and declaring yourself to be indigent – providing proof of your poverty.

Dawson makes the interesting observation that the meters had a debilitating political effect, isolating individuals from one another and turning the issue of payment or non-payment into an entirely individual matter – 'privatising', in other words, the very relationship with the state. From a discourse of solidarity, shared struggle and the public good, the ANC had moved to a language of 'entitlement' – in which rights are no longer undivided but conditional on 'responsibilities'. In other words, rights, and citizenship itself, had to be earned. And while these responsibilities fell on individuals and poor households, neither the majority users of water in South Africa – agriculture and industry – nor the wealthier (and far greedier) water consumers in the rich suburbs were required to pay in advance, nor were their tariffs higher because of their higher usage. For the APF, committed to the principle of the redistribution of wealth, and the fight against neo-liberalism, this was a deeply negative marker.

The response of the APF was direct action. In Soweto, the water meters were removed and households connected directly to the water pipes by groups of 'struggle plumbers'; in Phiri, residents refilled the trenches that had been dug in preparation for the laying of pipes. It is extremely significant that several of the companies involved in the South African privatisation, like Suez, have handed back their contracts because of lower than expected profits. And the director of Vivendi/Veolia, Yves Picard, made no bones about his company's reluctance to undertake work in South Africa because the poor do not pay.

Neo-liberalism works with universal models based on cost-benefit analyses and calculations of profit. It imposes these through its financial agencies and trade organisations irrespective of local conditions, and it imposes them without distinction on states. The tragedy of South Africa, its state mired as it is in corruption and power broking, is that its poor and working class, who identified with it throughout the anti-apartheid struggle, and flocked to vote for it when it was legalised, are now the victim of its policies. Nevertheless, in the process of struggle the more advanced social movements, like the APF, have developed both understandings and methods that enable it to act in the present but to organise as it does so for a very different, emancipatory, democracy in the future.

Ireland is currently in the throes of a mass movement against the Irish government's decision to impose water charges for the first time in its history through what is oddly described as a 'semi-state commercial company', Irish Water.[20] Like any other private water company, it is working for full cost recovery. There are parallels here with South Africa, particularly regarding water meters, which the government insists will be universally imposed. The meters themselves cost 539 million euros, far more than the anticipated returns from water charges. The previous system had a 50 per cent leakage right, but the new company has evaded any suggestion that its first priority is to repair them.

On the contrary, it has argued that most leaks occur within properties and therefore householders must bear the cost – indeed according to a government spokesman, they will be forced to do so.[21] Thus far, 66 per cent of potential consumers have registered – that is, just under one million households out of around one and a half million potential users.

The decision was taken in order to obtain loans to address the country's economic crisis, on the now familiar basis that privatisation would be a condition of any rescue package. But the response to the proposals has been enormous public discontent, particularly when the Chief Executive revealed that 50 million euros had been spent on consultants; that turned to rage when the real figure proved to be 86 million euros on consultants and 180 million euros on start-up costs. Privatisation of water in the UK had started in much the same way, with hand-outs to the water companies for consultancy and start-up costs. In October 2014, the first Right2Water protest brought 80,000 people to demonstrate in Dublin; in November, the march had swollen to 100,000. By then the installation of meters had begun, but the installers were met by angry groups of residents wherever they went. The government, impressed by the scale of opposition, then reduced the rates and capped them until January 2019 at 60 euros per one adult household and 160 euros for households of more than one. But the movement already has its martyrs – five people jailed, one for refusing to pay and four for incidents on demonstrations. Surprisingly they were released early by the High Court. Echoing the experience of the anti-poll tax movement which brought down the Thatcher government in Britain, the government is now threatening to deduct unpaid bills from wages and social security payments, starting in 2016, even though they recently claimed not to have sufficient funds for the calculation of rebates. As we write (in late March 2015), the movement continues to grow. But what the Irish government fails to understand is that what is being fought here is the loss of a public good, a resource.

The issue in Ireland is complex, since it involves paying for water, and we will return to it in Chapter 9. What is important, however, and the Irish government clearly does not understand this, is that what is at issue here is not the level of payments but the loss of a public good, a collective resource to which people have always felt a natural and unspoken entitlement.

Increasingly that is the issue around which water protests have been organised across the world. It is a serious analytical error to treat such movements as single-issue campaigns, unless democracy is also a single issue. Water, more than any other motive, embraces every aspect of social and political life – citizenship, rights, the distribution of wealth and power, the path and meaning of development, the concept of a decent, safe life, and the meaning of democracy. All this is in a single drop of water.

RESISTING WATER CHARGES
– IRELAND
Richard Boyd Barrett

After six years of crushing IMF-EU led austerity, the Irish people finally had enough and a mass movement came out on to the streets in the autumn of 2014 against the imposition of Water charges. The Right2Water campaign brought together community activists, trade unions and political parties in a series of national demonstrations through the centre of Dublin to oppose the charges and any proposal to privatise water in Ireland. On each occasion between 80,000 and 150,000 demonstrated – by far the largest demonstrations in Ireland in over fifty years.

According to the Government, the Irish people are being forced to pay for water because:

- It is a requirement under the EU Water Framework Directive;
- It is the only way to stop us 'wasting' water;
- Everyone else pays in the EU, so therefore so should we;
- Ireland has signed up to do so as a condition of the EU/IMF/ECB bailout package.

The truth is that we already pay for our water; we don't get it for free. We pay for it through our taxes. It may be true that most EU countries charge directly for water. But it is equally true that we pay high taxes for significantly less services. Irish Water is being set up on a 'Public Utility Model', 51 per cent state-owned, with the balance coming from private funds. As such it will be run as a commercial concern with shareholders seeking maximum returns and lenders demanding commercial interest rates. The public utility model is the most attractive proposition to lenders and points towards privatisation at some time in the future.

Why is all this happening, and what can be done about it?

Crony capitalism and 'jobs for the boys' have been features of Irish capitalism since Independence. But what is new is the immense debt and the policies of austerity that are destroying the Irish economy and society – a debt which was not generated by the Irish people. Debt now exceeds 150 per cent of Gross National Product. The Irish taxpayer is forced to pay €9 billion a year in interest payments. The government's aim is to make low- and middle-income people shoulder that burden – and let their wealthy friends off easily. Tax on the income of workers has gone from 27 per cent

▶

of revenue in 2007 to 42 per cent today, while tax on profits and capital gains has fallen as a proportion of the total tax take. Tax on big business is almost negligible. Water charges are, quite simply, just another way of making people pay for the economic crimes of the bankers, speculators and the wealthy. The history of water in Ireland since the 1970s is dominated by controversies about whether there should be charges for domestic supply. When domestic rates, which paid for our water and the treatment of waste water, were abolished in 1977, income tax was increased to compensate local authorities for the loss of revenue. But by 1983 charges were reintroduced by some local authorities. And not surprisingly income tax was not proportionally reduced. It was not until 1994 that some of the bigger cities including Dublin reintroduced charges. These were met with a campaign of boycott and resistance, leading to their abolition in 1997. Water charges were beaten before; we can do it again. The reality is that protest works. A broad campaign of street protests and civil disobedience is again required to encourage an awareness of the charges and their implications and to finally defeat them.

A strategy to beat the charges in Ireland?

The key to defeating water charges and water privatisation is to continue mass mobilisation at every possible level: mass demonstrations, community resistance to water meter installation, civil disobedience, mass non-registration with the new water utility, a national boycott of the charges, and the development of a radical left and anti-austerity united front, inside and outside the parliament. All of these tactics are being pursued by the anti-water charges movement. First, we mobilised mass people power on the streets to increase people's confidence that they can win. The campaign has worked to involve as many other supporters as possible, including local authority water workers and environmental activists. A broad alliance of protesters was mobilised of those directly affected by the charges and those who are shocked by the implications of the government's plans. This political revolt of protest and civil disobedience must continually be escalated to halt the rip-off of our water and its future privatisation. Such a campaign, if strong enough, could lead to a mass boycott. In April 2014 a small local community in County Cork mobilised to stop the installation of water meters; this then led to a series of communities organising to resist Irish Water in their estates. A number of areas in Dublin saw substantial attendances at protests and street meetings where everyone could voice their opinion and feel ownership of the campaign. This mass democracy was a vital part of the process of raising people's confidence and willingness

▶

9

What is to be Done?

We have set out to identify the many areas in which water, its provision, use, allocation and distribution, and to some extent its symbolic significance, are experienced and disputed in the world. The backdrop to our discussion, and indeed to most writing on water over the last 20 years, has been the notion of crisis, and its possible culmination in war. These are emotive terms, and dramatic ones – but they are also extremely non-specific and in the context in which the debate has arisen, serve to veil other more complex questions about responsibility and possible resolutions.

The context of this book and much of the other work we have consulted is globalisation as the expression of neo-liberalism. It is worth recalling that the free movement of capital to which neo-liberalism is devoted has implied the destruction of mechanisms of protection and sovereignty across the developing world. The very concept of development as an endogenous process implied a challenge to an unfettered capitalism free to appropriate resources, destroy resistance and export the products manufactured with those raw materials to markets in the countries from which they came. Dependency theory argued that development in the third world could only occur behind protective tariff walls and under the control of strong states able to shape the economy and distribute subsidies in ways that would encourage economic growth. That in its turn would provide the escape route from enforced underdevelopment, which was itself a function and a reflection of the development and growth of the industrial world. Development, in other words, was a qualitative as well as a quantitative process (as opposed to growth, which was simply the latter).

Some larger developing countries did achieve growth under these import-substitution regimes – like Argentina, Mexico and Brazil. But the response of economic imperialism was to use its economic and military power to destroy those experiments and re-impose the unrestricted domination of the market. Because the period of growth had strengthened the state and

to fight back. In areas where the meters had not yet
and the Right2Water Campaign called public n
networks of residents to fight. Activists in Ireland loc
of the water protests in Bolivia where mass protests, ,
and strikes beat the Government. The protests against th
water meters were only a starting point – a means to delay th
and inspire resistance. This allowed the campaign to build ,
residents in every area who could mobilise for the bigger natio,
which themselves offer the possibility of escalating into mass
obedience. The campaign against water charges will continue an,
process we also want to create a radical new movement for a better ʂ
that values and protects all its natural resources, including water.

Richard Boyd Barre,
TD, People Before Profit Alliance
and a member of the Irish parliament

the strength of workers organisation as industries expanded, the destruction of these national projects, which enjoyed mass support, was violent and sustained. Chile was emblematic in this respect. The brutal repression of the Popular Unity's project for social reform and economic development transferred power to a military regime charged with enforcing the priorities of the global market. The experience was repeated in Uruguay (1973), Argentina (1976), Colombia and elsewhere, and in Central America in the 1980s with devastating long-term consequences for all its member nations.

Yet this period was never described as one of crisis, but rather of restoration. And it emerged that it was a preparatory phase for a new assault by global capital, dismantling the public sector that had begun to emerge during the earlier period, in the name of an open neo-liberal ideology. It was 'liberal' in the sense of recognising no barriers to free trade and the free movement of capital, but it had very little to do with a classical liberal view of the relationship between individuals and the state, in which the state was the guarantor of inalienable individual social and political rights and the embodiment of a social contract. Marx, of course, described the state as 'the executive committee of the bourgeoisie', but in bourgeois democracies it had been represented as a body above class, and a representative of the interests of society as a whole.

The water crisis to which so many writers referred began to be discussed as globalisation embarked on its course of integrating the world market and imposing capital's laws of motion, unrestricted, on the planet. Development came to mean the centralisation of control over the world's resources and the systematic alienation of the producers, the workers. The dismantling of nation-states which began in the late 1980s replaced them with minimal state structures charged with controlling their own territories for capital, acting as its surrogate, and removing the legislation that offered some protections to the majority population in each of their territories. The dismantling of systems of social security, the assault on public health and education, was initiated wherever capital held sway, whether in the developed or underdeveloped world, giving free access to a marauding capitalism to seize hold of resources as instruments for its own growth; resources like oil, minerals, but also land and water. And its disciplinary instruments – the World Bank, the International Monetary Fund, the European Central Bank and the World Trade Organization – together imposed a global financial and commercial order.

But this also involved an ideological transformation. First, the creation of a global culture rooted in patterns of worldwide consumption served by new 'international' brands and products which could unify and homogenise social conducts. We became what we consumed, as Naomi Klein so dramatically and powerfully exposed in *No Logo*.[1] More importantly, the concept of social solidarity embedded in trade unions, working class and peasant organisations, was undermined and replaced by an ideology of competitive individualism reflected in a thousand television series setting individuals against one another in absurd invented scenarios. All this, of course, was made possible by the concentration of power in the mass media on a global stage, of which Rupert Murdoch is the embodiment. The cultural transformation can be described with the single word that has dominated discussions in the water field – privatisation. It defines not simply the appropriation of water resources by privately owned corporations, but also the assault on every expression of the collective experience. Trade unions, the welfare state in its many forms, a social system of health care, public education, each were undermined and replaced by private programmes with their accompanying ideology. And at the heart of it, the notion of the public good and the rights accruing to every citizen – the social contract – was no more. Instead, society became a market place in which citizenship rights were *earned*. The poor were divided into the deserving and the undeserving. The elderly became responsible for their own late-life care, while their careful savings for that moment in work-related pension schemes were taken from them and employed for purposes of speculation.

That same process was reflected in other areas, as major corporations were swallowed by larger and larger multinationals, the majority – at least until recently – based in the United States. Successive US presidents sustained and supported their expansion and their ruthless destruction of lesser rivals; the largest military and still the most powerful economic power in the world could and did do so with impunity. As global warming intensified through the 1980s, the power of these corporations was sufficient to call climate change into question and to delay the public debate of the measures that should be taken. The actions of the US government at each of the series of international environmental conferences, from Kyoto through Rio to Copenhagen showed very clearly that Big Oil and Big Coal and Big Water could rely on Washington to defend their interests come what may. And the absence of strong independent states, and the dismantling of their capacity to challenge dependency on the global market, made it increasingly unlikely that any of them would be in a position to challenge the hegemony of US imperialism. The confidence of that powerful capitalist class was well expressed in the document that laid the groundwork for the destruction of Iraq in the name of Big Oil – it was called 'The Project for a New American Century'.[2] The document was in a sense a response to the rise of other, capitalist centres which might at some point in a not too distant future be capable of challenging US hegemony – China in particular. India too was emerging as a major power, though not one that could contest the US domination of the world market – nonetheless the Indian corporation Mittal is the world's largest steel producer, while Pittsburgh like the other iconic centres of steel production in the United States now retains only one, rusting steelworks to evidence its past weight in the industrial world.[3] As environmental awareness grew in the face of the deterioration of the original municipal water infrastructures, the industrial processes that had caused the pollution and decay were also exported to a developing world where the restrictions increasingly imposed in the richer countries would not apply.

As we have argued, the World Trade Organization (WTO) drives the global strategies of global capital, unburdened by the inconvenience of international negotiation and conciliation which the United Nations, notionally at least, still represents. The WTO makes no obeisance to ideas of world peace or world harmony; it acts openly and remorselessly to further the interests of capital on a world stage. Privatisation, selling off national assets and public services, and perhaps most importantly systematically undermining the notion of social solidarity, has dismantled the welfare state and set out to destroy the ideology that underpinned it – helped, of course, by the desertion of many leading functionaries of those welfare states, tempted by the huge rewards available to those willing to carry through the auction sale of public goods.

But none of this has gone unchallenged. In some senses, water has come to represent and enshrine the whole process of globalisation. The real significance and implications of privatisation have been most dramatically exposed in the water field, though there is no area of public spending where it has not faced a similar challenge – health, education, social spending etc. The contempt for a decent, humane life is daily demonstrated in the great megacities where human waste flows past the giant towers of the modern age. The mindless destruction of the very material of survival by a 'reason-blind' system incapable of long-term vision is expressed in the scarred forest floors, the poisoned rivers, the dried out wetlands that are now the daily diet of features editors across the world.

That is why the struggle over water, the *politics of water*, has increasingly become the focus of public protest and collective resistance. At one level, it is because the lack of water, as the very stuff of life, can only have one outcome. A class society, divided between the rich and the poor and now redefined as those with or without water, will in the end inescapably lead to what Marx described as 'the mutual ruin of the contending classes'. Yet it seems that the race to dominate the global market, whatever its consequences, is all that matters to the rich and the powerful – who clearly see themselves as immortal. But for the world's majority, the victims of their madness, the realities come home daily in the form of floods, storms, contamination, sickness, hunger and thirst.

The arguments adduced to address the problem seem to rest on concepts of crisis, apocalyptic predictions of death and disaster from hunger or from the wars that the lack of water will provoke. But these persistent warnings lay the responsibility firmly at the door of nature (crisis as natural disaster) or of the poor themselves, in a veiled neo-Malthusian suggestion that the population explosion will be the cause of future conflicts, since there would be enough water if only ... Our purpose throughout this book has been to show and to repeat that the crisis is the product of human actors and human actions. This is not only to apportion responsibility, but also to say that by the same token, it is human intervention that can redeem the errors of the past and restore at least part of what has been destroyed.

The local and the global

Agriculture

Adopting sustainable agriculture does not necessarily mean a return to primitive farming systems.[4]

The green revolution and all that followed on from it was represented to the world as a means of alleviating world hunger. That continues to be the justification, sometimes implicit sometimes explicit, for industrialised agriculture. It is the explanation of why the amount of land devoted to agriculture has increased worldwide and why just under 70 per cent of the world's water is consumed in agriculture. But we know now that the law of diminishing returns operates in agriculture too. The Green Revolution (or revolutions, as we are persuaded that a second one is now under way, courtesy of Monsanto) did introduce new crop varieties that produced more, that were in some cases drought resistant. But those varieties consumed more water and were dependent on pesticides and herbicides, which became less effective over time. That led to new generations of weeds and insects that were resistant to the chemicals, and which weakened and contaminated the soil. The so-called Second Green Revolution will reproduce the same problems, but with one aggravating factor; the new varieties of seeds have been patented, in many cases, and local farmer-raised varieties forbidden under the arcane rules of the WTO. We have rehearsed these facts in earlier chapters. This not only ensures that pesticides, albeit of a new kind, will continue to flow through the irrigation channels of the developing world, contaminating just as their predecessors did. The farmers will continue to be indebted to Monsanto and its ilk as their meagre incomes will not cover the cost of the chemicals; yet they will be forced to use them,[5] and pay for them, and sink into the cycle of debt which is the main explanation for the huge numbers of suicides among small farmers in India who kill themselves rather than lose their land.[6] In any event, these debts are contracted mainly to grow cotton, an export crop whose water consumption is among the highest of any agricultural product. External investment, whether private or institutional, is invariably conditional on profitability – these institutions are after all lending institutions profiting from the interest on loans – and that is measured in the global market, where grain, maize, rice and cotton predominate. It is significant, for example, that while 5 per cent of India's agricultural land is given to cotton, it uses 45 per cent of the pesticides employed overall. The picture repeats itself across the world, whether in Peru – among the world's main producers of artichokes and asparagus, neither of which is part of the local diet – or in Brazil where maize, soya beans and grain are the principal crops, the first two mainly converted into bio-fuels (which are hugely water-intensive) and the last into cattle feed (which is the most water-exigent of all). It is unlikely that Kenya consumes the flowers and luxury vegetables grown in its fertile Rift Valley; and when drought struck the valley, those who cultivate them were left to suffer hunger. Maggie Black reports on the situation in the Senegal Valley, where rice-paddy cultivation was supposed to raise dietary levels. But their traditional foods were almost completely replaced by rice and a subsequent United States Agency for International Development (USAID) study found their nutritional intake had declined. They could no longer grow their traditional millet, maize and pulses because they had planted them seasonally on a flood plain that was destroyed by the scheme.[7]

The examples could be repeated ad infinitum, because modern irrigation systems are large scale and costly enterprises favouring industrial agriculture. And their impact on river basins, floodplains and wetlands has been amply illustrated across the world.

> In many parts of the world, growth in the rate of food production is slowing substantially or farm productivity is actually decreasing due to water shortages, desertification, increased soil salinity and other related factors. The per capita global production of cereals peaked in 1984 and has declined since.[8]

At the same time such food as is produced will increasingly go to the developed countries, which can afford to pay for them and divert their water to more lucrative uses – industrial or urban, for example. That inequality is reflected in what is grown in developing countries (see the example of Peru above), and in the poorest countries the fact that rural populations are growing food that does not feed them.

So what is the solution to what seems like a doomsday scenario in its more extreme expressions – the planet's population will grow, there will be a food and water shortage and the poorest will go hungry. The truth is that if the future were left entirely in the hands of the transnational corporations and the international financial institutions, that would be a plausible prediction. Feeding the hungry does not figure in the strategies of the food multinationals on the very simple grounds that the poor cannot pay. That is also the argument offered for installing prepayment water meters in Soweto and Detroit, rather than a concern that every member of society should have access to it as a matter of right.

Let's leave aside for the moment the statistical basis for the predictions of population growth, which assume that the rate of growth will not decline as the poorest societies change and develop, culturally as well as economically. The simple truth is that the industrial irrigation on which export agriculture depends is becoming less productive and less efficient, that more land will have to be devoted to growing crops to compensate for that declining productivity, and that on the other hand the growth of cities is absorbing farmland. 'Cleared tropical forests, for example, typically produce crops for only a few years until the soil is exhausted.'[9] It is easy to imagine that global corporations will simply cut more trees faster and pollute more soil in a desperate drive to maintain their profits.

There is general agreement that it is urgent that different methods are introduced which finally understand that universal solutions cannot address specific problems. Farmers know their soil and their local climate better than anyone else; they are the inheritors of a long collective memory embedded in traditions and local histories that are not mere distractions but enshrine practical solutions to recurrent problems. The mixed crop farming typical of the farmers of India is not simply a cultural habit; it is a way of sustaining an ecosystem over time; whether or not the individual farmers and their families know the function of each element, the point is it works. The 'modern' methods of farming have had the opposite effect – to interrupt and disturb the water cycle, so that rainfall and run-off do not simply seep into the soil, replenishing rivers and underground aquifers which in their turn water the plants whose evaporation becomes the rain, in a benevolent cycle that ensures the renewal of water. The process is interrupted now at every stage; desertification and salinity pollute the run-off, deforestation and dams contribute to the rising salinisation of waters, wetlands and rivers dry out so that flood plains and swamps remain only as parched, cracked and useless land.

There are alternative methods of irrigation whose impact on the amount of water and its productivity can be dramatic. Drip feed irrigation is fundamentally very simple; plants are fed at the roots by plastic tubes perforated to supply water to each root, but to avoid the rush of water down irrigation channels, much of which does not reach the plants at all but erodes the channels, contributing to the salinity of the soil and the rivers where the waters go. Sprinkler irrigation too has proven its worth, though not those whose arcs of water in the air fall randomly across the fields – these are ground level sprinklers whose jets can be directed. According to Sandra Postel, an acknowledged expert in irrigation and author of *Pillar of Sand:*

Can the Irrigation Miracle Last?,[10] drip irrigation can cut water use by 30–70 per cent and raise crop yields by 20–90 per cent – yet they are currently in use on only 1 per cent of the land; sprinklers, though not necessarily the updated kind, cover 15 per cent. Forty per cent of what remains is irrigated land. She details[11] a number of other methods, including monitoring and timing irrigation regimes more precisely and non-plough tillage which reduces dramatically emissions of CO_2 and methane from the soil. All of these methods, of course, could be employed by small farmers and local networks. A recent World Bank scheme to provide small farmers with mobile phones to which data on crop productivity and climate change are sent seems a good idea, but it will involve an expense that will once again marginalise the poor. Why are these methods not being implemented, given the clear evidence of significant water saving – and the urgent need? The answer will be found when we answer the question – who benefits? It is the big landowners who receive the bulk of subsidised irrigation water, and it is their cash crops that are watered and later exported. In Latin America, for example, it is clear that subsidies and technical support go to the 'modern' sector rather than the Andean farmers. It is the small farmers, who produce food for the rural populations, many of them removed from their land by the same deterioration of the soil and the drying out of rivers, who will pay the full price.

The Sahel provides a telling example of what this means in practice. For generations the different communities had coexisted in the arid area of Darfur, in Sudan. Some herded cattle, camels and goats; others farmed and kept domestic animals. In the dry season the herders travelled north in search of grass, and sometimes grazed their animals in the stubble of the fields or bought water from the farmers. It was a complex society, a mix (but not a fusion) of cultures, languages and identities, and there was an elaborate system of exchange that over the centuries had sustained all the communities in this harsh landscape. Then, through the 1960s, the rains failed as a direct result of global warming and the changing temperature of the oceans – and by 1983–84 this had reached famine proportions. After a period of civil war a peace was signed between the fertile, and much richer north of the country and the poor south. It lasted until the discovery of oil in the south when a national government based in the north found itself in economic difficulties because of the debts it had incurred with the IMF and others in the previous decade. Oil wealth beckoned and a new civil war began over control of the oilfields, where the American multinational Chevron was a dominant force. The combination of a war for oil and a climate change disaster turned Darfur into a byword for famine and suffering. When the rains returned two years later, the different communities of Darfur had become bitter enemies, based on what they had suffered during the time of disaster.[12] To the south, Lake Chad, once a mighty lake 25,000 square kilometres in size, has shrunk to 2,500 in less than 50 years. The communities living on its shores belong to four African nations; the disappearance of 90 per cent of the lake has meant that fishing has been replaced by farming, mostly growing food crops like corn, rice and sorghum. But there is a project to grow peppers and rice for export which will require pesticides and irrigation pumps. Both are costly and damaging to an environment that has already suffered from irrigation schemes in the past that contributed to the drying out of the lake and the desertification of the region. Yet in these unforgiving conditions, local communities have found ways to farm successfully with the means that they have available.

There are many examples. But where changes in irrigation methods could contribute to detaining, if not reversing the damage done by climate change and ill-considered large-scale irrigation schemes, the cost of drip feeder pipes and the technology to enable farmers to control crop rotation more closely is beyond the reach of some of the world's poorest communities. Yet we speak of cost as if it were an objective measure rather than a price charged in a market by manufacturers. It is a small example of how resources are denied to people by a market that considers only gain.

It is important to note, however, that these potential disasters do not only affect the poorest regions on earth. The contamination of North America's Great Lakes has reached disaster proportions, as 50–100 million tons of toxic waste are emptied into its waters each year by the oil refineries and nuclear facilities around its shores – and it is worth remembering that the lakes are a principal source of drinking water.[13] And it was only as a result of a devastating six-year drought that the Murray-Darling Basin in Australia, once the largest rice producing region in the southern hemisphere, introduced a strict water allocation scheme.

Crisis and control

The reality is that the global water crisis is in fact a gathering of many crises, each rooted in specific conditions and experiences. There is a crisis caused by the systematic mis-allocation of services and resources; a crisis of 'governance' as it has been called, that reflects the absence of agreed solutions and the unwillingness to search for them. It is in specific, local situations that they can be developed, but only when the global, neo-liberal order, which imposes universal solutions that express only the interests of the powerful heartlands of capital, be it the USA, Europe or China, has been replaced.

The process of globalisation, and with it the globalisation of water, has been made possible by an erosion of democracy – as Naomi Klein puts it, 'democracy has been traded away in exchange for foreign capital.'[14] As multinational financial institutions dominated by the rich countries and the multinational corporations have conditioned their loans and investments on privatisation, again particularly in the case of water, local and indeed national control over economic decisions has been sacrificed. In the process national states have been systematically transformed into agents of the global system, reflecting and reproducing its ideology and its priorities. The call for democracy that recurs in every local struggle, therefore, is about the location of the decision-making process.

The symbolism of water is particularly revealing of the processes.[15] In most cultures the flow of water is symbolic of time itself, its free movement the expression of change and progress. The water cycle is the guarantee of renewal. Its interruption places the future development of mankind at risk. But here 'development' has a significantly different meaning from the one attributed to it in the purely economic language of the market, where development is simply growth, rather than the integrated evolution of humans and their environment in the progress towards a fulfilled and secure life.

It is neither science fiction nor primitivist fantasy to recognise that the ecosystem is not nature separated from man. It is a single organic whole of which human beings are a part. The disruption of the relationship of its parts, threatens the whole.

Pollution

According to the Blacksmith Institute and Green Cross International's annual report on pollution it is 'pollution – not disease – [that] is the biggest killer of children in low and middle income countries.'[16] This should be seen together with the figures for water-borne disease, but the point is well made. In poorer countries death comes from toxic waste from industry and agriculture, from the recycling of lead batteries and electronic equipment in Agbogbloshie in Ghana (which has the distinction of beating Chernobyl to the status of the world's most polluted place), from the mining that releases millions of tons of mercury from gold (artisanal and industrial) into the rivers of the Amazon or the copper mining whose cadmium and other heavy metals in rivers flow into the sea to reappear in fish thousands of miles away in the Arctic. China is the world's worst polluter, since its environmental controls are limited in scope and rarely function. A recently announced law to fine or even imprison polluters is unlikely to dissuade the mammoth industries of China from producing, given the profits to be made and the paltry fines they risk. While the USA allocates $8 billion per year to its Environmental Protection Agency, Blacksmith estimates at a few hundred thousand dollars that most African countries *might* have available.

Western governments have therefore applied controls and raised budgets for environmental protection, but this does not include limiting oil and coal production nor making the polluters responsible past and future. The costs come from a public purse which is paid for by every citizen, though the profits gathered from these activities remain with the global corporations who are responsible. In 2010, the UN agreed to the Aichi Biodiversity Targets at a cost of between $150 billion and $440 billion per year globally. These figures tend to lose meaning, but compare that with the USA's military spending in 2014 of $581 billion (4% of GDP) or China's for the same year of $130 billion (2.1% of GDP).[17]

As the air warms with global warming, the broader effect of pollution which contributes directly to greenhouse gas emissions, air pollution will increase too as stagnant air accumulates soot, dust and ozone. This is not intended as a nightmare scenario, but as the issues that have to be addressed by governments and social movements. Like the need to recover the soil and water that will grow our future crops, here too the manifestations of the problem of pollution will be local, even though its source is global; and it will be locally where it will first be addressed and fought. It is the aggregation of these local struggles that can and will become a global movement, imposing controls and discipline on an economy whose prevailing forces will never, of themselves, confront these consequences. Blacksmith is clear on the point: 'It is impossible to have healthy, properly functioning wetlands, forests or other green infrastructure if they are exposed to high levels of heavy metals, obsolete pesticides or radioactive waste.'[18]

Recycling

Recycling has certainly become more fashionable in the West. Where I live, in Scotland, there are queues at the recycling centre on Sundays. It is almost as if it has become an alternative to Sunday in the park or the stroll around the local IKEA. It is easy to dismiss this as tokenistic, but it is a small sign of awareness that the global movement can build on. Yet the destination of all this recycled stuff may be less obvious to people. A high proportion goes to landfill, with no guarantee that it will not, in time, contaminate that land. Some goes to reuse, especially glass and cardboard, which does have an environmental impact. But much of the physical material that is recycled will end its life in Africa and Asia, where it will disgorge its contaminants into a different environment, as we have discussed. There are two consequences that flow from that. One is to increase the pressure for the lowest energy treatment systems. The other, which is more long term and a more profound issue, is to address where that waste comes from in the first place. Household waste is probably largely packaging in one form or another, much of it the plastic that is a by-product of oil. Local campaigns to stop using plastic bags may be a small sign of things to come.

But by far the more urgent issue is the recycling of wastewater. The UN's figures on those without adequate sanitation (2.6 million) is almost certainly an underestimate, because while it may assess whether people have toilet *facilities* it does not address the mammoth problem of where the flush takes the water. In fact human waste as well as industrial and agricultural waste flow largely untreated outside the wealthy countries into sewers or irrigation channels or pipes that take them straight into rivers and waterways. Yet wastewater, or 'grey water' as it is called, could be recycled and reused for non-drinking domestic purposes, for industrial cooling for example, or for irrigation. Singapore has been exploring these possibilities for some time, but as membrane technology advanced, it was able from 2003 onwards to recycle wastewater for these purposes and ultimately to achieve drinking quality from it. It also charged the full cost of recovery (a controversial issue we shall return to) but the costs to consumers fell by about 5 per cent by 2014.[19] This is not to argue the case for pricing water, but simply to show that contrary to the arguments coming from the private sector, recycling can not only conserve water – it can be reused, and it need not imply rising costs. We have discussed earlier the very high levels of leakage and loss in almost all the world's cities; there is no reason why urban infrastructure cannot be properly maintained, as Tokyo and Phnom Penh have done. And to anticipate a typical response, privatisation of water is not a solution. Most private water corporations are interested only in the distribution of water and in water services; the presumption invariably is that infrastructure will remain a municipal responsibility. That is why private corporations withdrew from water contracts in South Africa, Manila, Delhi and elsewhere on the grounds of insufficient profitability.

Neo-liberalism has circulated a number of myths about municipal ownership and management of water; they are part of a wider vocabulary discrediting all forms of public ownership and control.[20] In fact many municipal water companies were efficient and professionally run, but it is probably true that some were not, and that they deteriorated in the last two decades of the twentieth century, the period in which the systematic cutting back of public sector budgets began and when privatisation began its quiet intrusion. It is also true, and especially so in the developing world, that state-run institutions are prone to corruption and all that goes with it. In part, that has to do with the role that the state has come to play in the neo-liberal era, as the agent of multinational capital and the global financial institutions.[21] The extent to which that is true is in direct proportion to the level of democratic accountability in the society; the greater the genuine involvement of society as a whole, the more transparent government is, and the more an authentic living democracy can take the place of the democracy that is limited to a periodic vote for representatives chosen at a distance from the grassroots, the more corruption and mismanagement will be addressed. There are some examples to be pointed to to demonstrate the possibility of turning the tide in that sense; the processes that carried Evo Morales to power in Bolivia and Hugo Chavez in Venezuela, the election of Syriza in Greece, the growth of Podemos in Spain all point to a possibility of change.

And in the specific context of water, there is no technological or economic reason for the collapse of many urban water systems, the high proportion of leaks, the contamination of public water supplies.[22] Though water supplies in urban settings are clearly under pressure, repair of infrastructure (compare Tokyo's figures for leaks with, say, Mexico City's or even London's)[23] is clearly possible. Recycling, as Singapore shows, is also feasible with current technologies. And there are other technologies – desalination may be one, though it is demanding of fuel and currently expensive. Nonetheless one-quarter of Barcelona's consumers are served by recycled water, as are 30 per cent of Israel's. But while seeking out alternative methods of recycling and of mining water, the priority should be revisiting the way in which existing water supplies are managed. The contamination of water sources remains a central concern; potable water comes from the same rivers and aquifers that are polluted by human waste, agriculture and industry. Municipal management is no better or worse than private management *in itself*;[24] the difference is that municipal utilities are, or should be, publicly accountable at every level. They should be run, as every utility should be, by professionals in the field, rather than by political appointees prone to the pressures of patronage and clientilism. And they should be run according to an ideology of public provision and service. This may provoke sceptical noises, but why are the founders of the public health regimes in the West different from those running the service now? They are not a different breed, nor is there something inherently better about nineteenth-century intellectuals as compared to twentieth-century ones. It is a matter of the ideological framework within which these services exist. And let us, as socialists, be honest about this; in the majority of cases their ideology was social-democratic, perhaps even paternalistic. They worked within a capitalist framework, but one in which the organisations of working people had won rights and services in exchange for a temporary compact with capitalism. It was certainly a vastly more benevolent system than the ruthless savage capitalism of the globalised world today. It benefited some and it consigned others to poverty that is true. We are not defending social democracy, but pointing to something very different. That capitalism was capable of that compromise in a time of expansion and economic growth. What we see now is capitalism in crisis and under pressure, with its gloves removed and its true brutal nature exposed. That we should be arguing now, against the imperatives of capital, for the just distribution of that which is life itself – water – is the evidence of how savage capitalism is. In a sense, the crisis of water is a crisis of management, as Professor Biswas and others describe it – but we should understand management as the values on which society and the distribution of wealth and resources is decided.

It will be argued that water can no longer be simply taken from rivers or lakes, as it once was. That is true of course, not least because so many of those sources have shrunk, disappeared or become contaminated. Water in the twenty-first century has to be engineered – treated, measured, stored and transported; 'God gave us the water, but not the pipes', as one Suez executive is reported as saying. We cannot answer for the Grand Design, but water was not designated to wealthy areas nor was sanitation meant only for the few. Sadly, the engineering miracles of the modern age have been disproportionately assigned to military purposes and we have been left with Teflon frying pans.[25] If simple regulation of water quality were rigorously implemented worldwide, and the technology supplied *without cost* to the developing countries of the world we would be discussing a very different situation today, both as regards provision and water quality. In the medium term, there is a growing campaign to restore rivers and lakes. This is neither cheap nor easy, but it can and should be done. Sandra Postel, the irrigation expert, heads a campaign to restore water to the dried out Colorado River, the water artery of the American south-west. It is an important and valuable idea, but it is hardly convincing when water misusers like Coca-Cola are backing the campaign, and claiming it for their own. This is, of course, a key question. Corporations trying to win green credentials (and new customers) are, in our view, very unreliable allies in the fight to reclaim water, which has been polluted and misallocated by those corporations themselves. The Conclusion to the UN World Water Development Report 2015, for example, offers some thoughtful responses to the world water crisis but in invoking the support of 'business', presumably banks and the corporations they finance, it undermines its general position of sustainability and transparency and the participation of all sections of society.

There is a dangerous argument that private corporations are in a better position to address environmental problems than public utilities – it stems from the assumption we referred to earlier that the public sector is inefficient, corrupt and so on. This, of course, is an ideological position not a scientific finding and there is plenty of evidence of the corruption and inefficiency of private capital. The claim to environmental credentials by corporations like BP, Coca-Cola and others is cynical in the extreme. And that will become clear as soon as the democratic content of a genuine movement for the restoration of public goods is established.

Forests can be restored and replanted, soil can be replenished in a relatively short time, rivers can be returned to their course, lakes restocked – but there will be no quick return on that investment, which is why private capital cannot and will not do it and why no bank or financial agency will invest in it.

An economic good?

The 1992 International Conference on Water and Environment in Dublin concluded with a declaration of four principles: 1) water is a finite and vulnerable resource; 2) water development and management should be based on a participatory approach; 3) women have a key role to play in relation to water; and 4) water is an economic good. What is significant here is the absence of any reference to water as a public good or a right and the supporting assertion that water is finite. Of course it is, but unlike oil, it is renewable – or it can be, as long as the water cycle is not disrupted or contaminated. But that assumption leads in turn to the assertion that it is an economic good.

As we have argued elsewhere, an economic good is one whose value is established in and by the market, a commodity like any other.[26] This leads to a pervasive argument coming largely from neo-liberals that conservation can only result from making water scarce or unavailable – at least to some. In a capitalist world market that simply means that those who can afford to pay, will pay, and those who cannot will go without. We have already seen that in many parts of the world the poor pay far higher prices for their water than the wealthy, and that the proportion of their income spent on water (it is generally agreed this should not be more than 2–3% at most) can reach 20 per cent. That means of course that they will have to choose, as many people did when Thatcher privatised water, between washing and drinking, between flushing the toilet and taking a bath, between drinking and eating. The World Bank has argued for subsidising water for the poor; this perpetuates poverty and demands and humiliates those who cannot pay, as we saw in Ireland and South Africa. But the crucial point is that the argument for pricing water refers only to the individual consumer (the 10% consumed domestically) and not to the corporate interests who should be paying for the water they use and profit from, but do not. Maude Barlow and others argue for a tax on financial speculation, but that seems illusory in the wake of the financial crisis of 2008.

The second aspect is the implicit suggestion that water is an economic good *as opposed to a public good to which every human being has access as of right.* Making water available only to those who pay – or as in the South African case additionally to those who are willing to undergo a public humiliation to get it – assumes that water is owned by those who distribute it. It is not and cannot be owned in that sense, because it is a commons. What then about the argument linking price and scarcity. It is society's obligation to provide the necessities of life, and to organise their distribution according to the rules of social justice; if there is a cost to ensuring them then that is a cost that must be shared across society and paid for by a tax system that is equitable and redistributive. In a word, access to water is an issue of democracy. An authentic democratic system will distribute resources according to need; our current order distributes them according to the power of the applicants and their economic weight, precisely because it is a commodity regulated by price. That is the position that the Irish Right2Water Movement is fighting today.

In a different kind of society, based on the public good, there may be an argument that people should contribute towards their water needs above the necessary life-sustaining provision, based strictly on ability to pay on the basis that no one should pay above 1–2 per cent of their income, and that industry and agriculture should pay a full and realistic tax based on their water use. This is an issue that is still controversial, though both Trevor Ngwane of the South African Anti-Privatisation Forum and Professor Asit K. Biswas argue this. But while we would regard water provision as a right and a responsibility of the state, the history of state provision has been fraught with failure and mismanagement across the planet – with some noble exceptions. We have argued that a state that is permanently subject to pressure from global capital and its market cannot be relied upon to introduce a genuine democracy, with control and government in the hands of the grassroots. In David Harvey's words:

> We can interpret neoliberalization either as a *utopian* project to realize a theoretical design for the reorganization of international capitalism or as a *political* project to re-establish the conditions for capital accumulation and to restore the power of economic elites.[27]

Everything we have discussed and described in this book leaves no doubt that it is the latter.

For the moment, movements from below will have to battle with states and government and organise themselves accordingly. But pollution and contamination, seriously endangering the water base for all, cannot be simply addressed by fines (which those who pollute can well afford and which in many cases will never be paid) or jailing the occasional executive, given that others will appear to continue the polluting. Contamination is effectively a crime against humanity which needs to be dealt with the utmost severity – with enforced closure as an initial step.

10

A New World Water Order

As the world has become aware that water is in every sense in danger, the reactions and resistances have spread across the world. The nature of the movement, its varied forms and expressions, is emblematic of the way in which a New International Water Order can be created. In 2015 we can justifiably speak of a global movement formed of a thousand smaller struggles and resistances rooted in their time and place. That diversity is its strength. But the movement has also created alternatives, examples of how to treasure and conserve water and live in a harmonious relationship with our surroundings. In fact, it is wrong to speak about surroundings, as if we were separate from the environment in which we live. We are a part of it, and we would do well to remember that, even as we are harnessing nature to human purposes.

Reflecting back on the debate to which this is a contribution, it is striking how long ago the problem of water provision, scarcity and conservation was posed. We have not looked as far back as the unsung hero Major Archer, who successfully reforested north-eastern Brazil at the behest of the Emperor Dom Pedro II in the nineteenth century, nor given adequate credit perhaps to the pioneers in the field, like Dr John Snow and the pioneers of public health. But for some time now, a number of far-sighted people have been warning of the transformations the world was undergoing. Chico Mendes, who campaigned for the Amazon forest was murdered by loggers; Vandana Shiva has been a forceful voice on behalf of millions of small farmers; and Sabino Romero, who led the indigenous movement in Venezuela, was murdered as he fought in defence of traditional territories against the Venezuelan state mining giant Carbozulia. And as the early tragedies that were the product of global warming, the droughts and floods that have become familiar phenomena, a growing number of voices were raised to alert the world as to what was about to happen. It was a battle to be heard, of course, over the insistent boom of neo-liberalism's advocates. But in the end, what they lacked in power they made up for in numbers.

In 1977, the United Nations Water Conference, the first of its kind, was held in Mar del Plata, Argentina. Its aim was:

> to promote a level of preparedness nationally and internationally which would help the world avoid a water crisis of global dimensions *by the end of the present century*. It was to deal with the problem of ensuring that the world had an adequate supply of water, of good quality, to meet the needs of a global population which is not only growing but also seeking improved economic and social conditions for all.[1]

The conference ended by launching an International Drinking Water Supply and Sanitation Decade to begin in 1981. Looking back ten years later, one of its participants, Professor Asit K. Biswas, professed himself 'disheartened' by what he found. There had been some movement, but by and large those who were intended as the beneficiaries of the decade continued to lack the resources that could improve their social and economic conditions. Today it seems that many of the arguments presented at Mar del Plata have been repeated again and again in the intervening years, though with an increasing sense of urgency. The Millennium Development Goals for water, agreed in 2004 for fulfilment in 2015, seem to have been significantly more modest and at serious risk of non-completion as we write.[2]

Why did the ambitious goals set out at Mar del Plata not move the world community to action at the time? Perhaps the assumptions of future crises seemed distant and unreal then; or perhaps the spokespeople of the emerging neo-liberal global order (this was just four years after the military coup in Chile) were able to mobilise their huge resources to undermine the conference conclusions, as they did with global warming. The continuing debate has been dominated, as we have tried to show, by international bodies and governments who are the prisoners of the multinational corporations whose priorities (and whose politics) have changed the very nature of the debate. When a group of concerned scientists formed the Intergovernmental Panel on Climate Change (IPCC) in 1987, they were subjected to a sustained and systematic campaign of discredit and denial, which certainly set back the discussion of climate change, and its impact on water, by a decade or more. In the age of globalisation the depletion of the planet's water resources has accelerated, and the divide between rich and poor, both within societies and between nations at a planetary level, has widened. In the interim, wetlands have been disappearing, rivers drying, forests shrinking, global temperatures rising. Meanwhile the unequal trade between North and South grew to the benefit of the multinational corporations that were increasingly shaping world governance, through the World Trade Organization, the World Bank, the IMF and others.

For many of the voices that now began to be raised in Defence of Water for Life, as the new UN Decade was called, the symbolism of the water cycle encapsulated the political, cultural and ideological impulse behind the new campaigns. For while the policy of Integrated Water Resources Management was emerging as a broad global response to what now was insistently called 'the water crisis', the reality of water systems in the context of an aggressive global capitalism was the disruption, indeed disintegration of the water cycle. The cycle that ensured the renewal of water was a perfect example of an ecosystem, in which each element interacted with every other – water, soil, flora, fauna, forests, air, organic matter, water vapour and so on. Yet pollution from agriculture and industry and from the open sewers that carried human waste into the system, reduced the sum of clean water available for drinking and washing; the over-pumping of aquifers reduced the recharge rate and compromised the cycle and the future supplies of water; global warming attacked the phytoplankton on which the ocean's fish stocks fed and began to melt the glaciers and ice caps that renewed the great rivers flowing down from the Andes and the Himalayas. And the pesticides and herbicides that fed export agriculture not only contributed to the heating of the atmosphere but also poisoned the soils where once food had grown.

The ecosystem is not simply nature, however, but men and women *in* nature; we too are part of the ecosystem that was being destroyed in the ruthless and blind pursuit of profits. How could the process be stopped? What was the alternative human order in which the protection and stewardship of the ecosystem of which we ourselves were a part could avert the crises and conflicts that the predicted thirst and hunger could bring? And these were not fantasies; one million people died in Papua New Guinea in 1997 as a result of drought. Such tragedies were increasingly repeated as the twenty-first century initiated the 'silent tsunami' of which Vandana Shiva speaks. What were the methods and actions that could reverse the process? What would a World Water Order based on cooperation, solidarity and an ecological awareness look like?

There is no blueprint for that other future; it will be different from all that went before because it will emerge from a democracy of subjects shaping their own history out of their experience and consciousness. It is a future that will have to be fought for. There are already living movements building that future, step by step, as they resist the depredations of neo-liberalism and forge new instruments of democratic control of public goods and services, and above all of water. A new International World Water Order will grow out of their experiences and their visions.

Steps towards a new world water order

The growing number of water movements around the world are successful precisely because they begin from specific experiences and organise around them. The demands of water movements in the cities of the USA, which are increasing in number, will of necessity be very different from the aims of movements in the Andean highlands or the flood plains of Pakistan and India. That is the strength of the movement, and the reason it cannot be subsumed under global programmes that become abstract in their distance from the grassroots. Nonetheless, there are some general principles that can be drawn from all those struggles and applied to strengthen and develop them individually and collectively. We suggest below some of those general themes and how they might be addressed in the immediate future and in the longer term.

Some principles for a world where the water flows free and plentiful

1. Water is the basis of life, a common resource that belongs to all living beings as of right.
2. Every human being has an inalienable right to clean drinking water and adequate water for hygiene and sanitation.
3. Because water is a fundamental human right, a public good, it cannot be alienated or privatised.

4. Therefore water cannot be a commodity, defined in its sale and purchase.

5. Water is renewed through the water cycle, an ecosystem that must be protected against misuse, misappropriation, and contamination by chemicals, pesticides, metals or other poisons that are the product of industry or intensive agriculture.

6. Water shall not be traded. Where there is an urgent need, water can be shared on a short-term basis.

7. The damage done by water projects driven by private profit and the domination of poor regions by the richer countries must be reversed as a matter of urgency.

8. There should be an immediate moratorium on dam building, and an emergency programme for the restoration of wetlands and flood plains and the return of rivers to their original course.

9. Future water projects shall be financed by public-to-public lending, and on the basis of cooperation and/or equal exchange. Banks and corporations, including the World Bank as at presently constituted, should not be permitted to influence decisions regarding water administration in any country.

10. All future environmental decisions should be taken in open local public assemblies and all research and documentation be open and available to all parties.

11. Scientific research shall be publicly funded and open to public scrutiny; commercial secrecy may not be a criterion for withholding information.

12. Local knowledge and water administration arrangements may not be overridden by central decisions until and unless they are approved in a democratic assembly of those affected.

13. It shall be an obligation of government to provide a complete municipal water service which is publicly run and cannot be sold to private interests. Such utilities shall be run by elected committees employing skilled professionals. They shall be regularly audited by elected grass-roots bodies.

14. Urgent action to be taken on emissions. These shall not be fines imposed on polluters but the immediate closure of polluting industries and crops and a ban on sales of polluting agents in industry or agriculture.

15. Systematic public education campaigns shall be organised to encourage people to consider their own 'water footprint'. This shall not be a

coercive process. Food and drink advertising should be banned or very carefully monitored so that it makes no false claims. Bottled water should be phased out wherever possible as soon as public utilities can provide assurance of the quality of water they are providing.

1. Water as the basis of life

There are very few cultures that do not have this concept at their heart. For a modern age, however, it is important to reiterate it. It is a principle that clearly distinguishes water from any other resource, in particular from fossil fuels, for example, which are owned and controlled worldwide by a handful of huge multinationals and a handful of oil-rich states, all of whom market oil as a commodity. The central battle today is to re-establish that water is what drives the complex system of which all human beings are also a part. It is therefore a 'commons' or a 'public good' which cannot be appropriated or privatised, any more than the air we breathe. It is obvious, yet in recent decades water has been redefined, in principle and in practice, as a market good, or commodity, to which access is determined by economic capacity. The first principle of any movement rooted in democracy and social justice, as all water movements must be, is that water cannot be privatised.

2. Against privatisation

The reality is that millions of people are still denied direct access to safe, publicly supplied drinking water and deprived of adequate sanitation. The result is that the poorest people in the world pay more than anyone else, both in terms of price and as a proportion of their income, for water whose sources are often unclear and which is often contaminated. In fact, even those with better incomes who are able to buy the billions of bottles of proprietary water that are sold every year, are often buying a contaminated product subject to less regulation than water distributed by public utilities.

The bottled water trade is one aspect of privatisation. But water is being privatised in many more ways. The water trade is estimated to be currently worth close to $50 billion, most of which goes to three or four multinational water companies who have over the last 30 years or so come to control water services across the world. Statistically, public water companies are still the majority, but in a neo-liberal world many if not most of their functions are contracted out; while the public utilities are starved of funds, banks and international lending agencies will provide funds to pay private corporations. In Britain water was privatised under Margaret Thatcher in 1989; the result was higher prices for water and lower standards, and that has been the pattern wherever water has been privatised. A service was thus transformed into business, and the availability of water was determined by how much you could pay. That is a scandal and an offence against social justice.

The availability of water can never be determined by income. It is a basic right of all human beings to have safe water to drink and clean water to wash in. It is society's obligation to provide both, and the job of governments to ensure that it does. It is unacceptable that the global financial regime obliges government to hand over the water sector to corporate interests. Privatisation is an attack on water as a public good. We must restore the public utilities, ensure that they are run by professionals and are accountable to society as a whole. There should be elected water committees in every area who will decide in public meetings how local water should be best used and allocated, ensuring that industry and industrial agriculture pay adequately for their water usage. And the water infrastructure, which is leaky and out-dated almost everywhere, should be addressed and repaired as a matter of urgency.

3. Dams, rivers and wetlands

An ecosystem is not something outside social life and human activity, but the foundation of our material existence. Water is a right, but it is also a necessity; so too is the material world that is sustained through the water cycle, the constant renewal of soil, rivers, seas and forests and the living beings that they sustain. Any interruption of that cycle threatens the whole interconnected and mutually dependent system in which each part is indispensable to the whole. It is, to use Marx's term, a 'totality'. Yet that disruption is happening and increasing in scale and speed.

Big dams were an expression of a global drive for growth in the richer countries. They were engineering projects first and foremost, competing in size and the amount of water they stored. Like all massive water projects, they were driven by money. Their effects did not become immediately obvious, but by the time the USA stopped building them in the 1970s and then began to dismantle them, it was obvious that their impact on the environment was devastating. Their negative impact was later confirmed by the World Commission on Dams reporting in 2000. The numbers of people displaced by dams are counted in tens of millions, the flood plains and wetlands drained by the arrested course of rivers are widespread, the methane gas released by the rotting plants at reservoir surfaces and the CO_2 emitted by the poisonous algae on the lake beds is a major contributor to global warming. It is now recognised that dams do enormous damage, and that the dried out rivers and wetlands have extensive and disastrous human consequences.

Beyond that, industrial pollution, the contamination of rivers and waterways by pesticides, herbicides, the dispersal of mercury and heavy metals from mining operations, the vast amounts of water used in oil production and especially in fracking which cannot be returned to the water cycle, all need to be immediately regulated and harshly controlled. They are the major cause of global warning through the emissions of CO_2 and other greenhouse gases, as well as the source of pollution and environmental contamination.

Governments can regulate greenhouse gas emissions if they choose to. In the USA and Europe there is increasing regulation and oversight to both limit and control emissions. This is a small step and an insufficient one. There has to be an active commitment to developing alternative energies – solar, wind etc. – and to a reduction in current uses of energy. The issue is too urgent to be postponed yet again or governed by strategies too long term to be of any purpose. In the developing world the situation is far worse; India has environmental legislation more often observed in the breach by governments under pressure to compete in an unforgiving world market, while in China, the second largest polluter after the USA, and probably the world's most polluted nation, very small steps are now being taken in the face of environmental protests. Governments will only act under pressure of public demand. There should be an immediate moratorium on dam building, and serious programmes for rewatering lakes and wetlands (as is currently being developed for example for Lake Chad) and for restoring

rivers to their original courses, as is beginning to happen with the Colorado.
This is especially urgent in India.

4. The banks, the WTO and the markets

Globalisation is the end point of a process of neo-liberalisation of the world economy. Neo-liberalism is the ultimate expression of an idea of free trade that is only free for some and very unfree for the rest. Milton Friedman, the author of the theory, saw the free movement of capital and its domination of the market as the ultimate expression of capitalism's driving impulse, which Marx encapsulated as 'Accumulate, accumulate! That is Moses and the Prophets!'. While capital and its most powerful actors, the multinational corporations, hid behind the protective shield of the US military and its government and state, other national measures were unacceptable 'restraints of trade'. Subsidies to local producers, for example, or tariff barriers to protect them, were unfair barriers to trade, and the organisation of working people to protect the value of their wages and their conditions of work were also obstacles to the free movement of capital. While the corporations and the banks could move their resources at will in search of profits, in pursuit of the cheapest labour or local resources, any obstacle placed in their way was an unfair limitation on free trade. Thus for example, Monsanto, the giant chemical company, can impose its genetically engineered seeds on small famers in India and then use its mighty managerial apparatus to persecute any farmer that uses their own. It is always an unequal battle in which the poor lose.

Globalisation has signified the sweeping aside of these so-called obstacles, by any means necessary – beginning with the military coup against the Popular Unity government in Chile on 11 September 1973. Its offence? Attempting to implement a programme of economic and social reforms for the benefit of the Chilean population, most significantly nationalising the copper mines which had provided huge profits to two US-based mining corporations, Anaconda and Kennecott. In the global market, where the multinational corporations compete to dominate and control, the banks and the international financial institutions have been indispensable allies.

Water projects – dams, irrigation schemes, for example – tend to be expensive. When governments or institutions apply for World Bank or IMF loans, they are conditional on being invested in for-profit activities, and often with full profit guarantees. These are usually with enterprises based in the rich countries who are paid directly by the banks; it is national governments who must then repay the loans with money from the public budget. These schemes are often described as Public–Private Partnerships (PPPs) but they are 'partnerships' between the poor and the rich and thus profoundly and enduringly unequal. It is a process that has been aptly described as 'the privatisation of the state'.

This inequality is then reinforced by trade regulations which reproduce and legitimise that relationship. The World Trade Organization, in effect the governing organisation of the world market, imposes rules to favour the most powerful and penalises and sanctions weaker states that attempt to limit corporate interventions. This now extends to services and 'intellectual property rights' – in other words knowledge itself, which is patented (i.e. privatised) by the corporations – through GATS and TRIPS respectively. And where corporations are refused access to natural resources, they may then take the country concerned to the International Centre for Settlement of Investment Disputes (ICSID), which is effectively run by the corporations and can impose huge fines.

The fraudulent Public–Private Partnerships benefit only the financial and industrial institutions of the global market, and tie their client states not only to accepting private investments in the most unfavourable conditions, but also determine where the money should be invested – not in schemes that are socially necessary, for example, but in profit-maximising strategies. This structure should be replaced by interest-free loans and cooperation between public bodies (which are called Public–Public Partnerships [PUPs]), who will have no influence over where the investments are placed. It is crucial that these financial arrangements are transparent, open to public examination and publicly audited, to avoid the corruption and bribery so characteristic of private investment in the public sector.

There are, of course, other international organisations, like the United Nations. But the reality is that the UN is constantly subordinated to global corporate interests. It is significant, for example, that 40 countries did not support the resolution on water as a human right at the UN General Assembly when it was proposed by Bolivia in 2010. These included the USA and China, as well as Canada and most European countries. The USA has also failed to adopt the United Nations Convention on the Law of the Sea (UNCLOS) and has refused to sign any of the declarations of the Earth Summits from Rio to Copenhagen. These facts could not be more eloquent.

5. The local and the global

A global movement in defence of water rights, and of the environment, is already growing. The way it grows and the relationship between its parts might be seen as analogous to the water cycle. It is an aggregation of its many parts, each contributing to its own immediate environment, its microsystem as it were, but also to the whole complex system. It is a movement in the sense that it represents a growing number of actions across the globe, but not in the sense that it marches behind a single banner. On the contrary, its demands are democratic because they rise from the bottom up.[3] Water scarcity, poverty, choked reservoirs, dry rivers, are lived in particular places and specific cultural conditions. Each part of the struggle in that sense is necessarily local, in that we aspire to a movement which draws more and people into a direct battle to govern their own lives. It is perfectly right to pressure governments and institutions, in the infinite number of small steps that will bring about global change. But it is important to recognise too that there is no example of government action for democracy, social justice and environmental improvement that has not been a response to mass activity and mobilisation. In a period where social democracy prevailed, capitalism was more inclined to accede to democratic demands because its priority was to maintain production in conditions of consensus. Under globalisation, however, there is no such disposition; the *politics* of globalisation are global domination, whatever the human cost. Every movement for social justice, environmental improvement and redistribution is a counter to the politics of savage capitalism.

What then is the politics of the water movement. Its watchword is democracy, but a democracy of genuine participation. The *UN World Water Development Report* of 2015, like those before it, lays great stress on the concept of participation. But there is the kind of participation which involves important people announcing their decisions to a selected audience, and there is the democracy of mass involvement, activity, engaging growing numbers of people in their own battles, and democratising the knowledges that will enable them to resist what Ricardo Petrella has called the 'mythification' of the reality of water, its uses and distribution.

Every victory however small, adds to collective confidence that change is possible as well as necessary. And every victory, however small, strikes back at neo-liberalism – an ideology that is built on the assumption that the vast majority of mankind can do very little against their power and wealth.

Water wars or water peace

The widely quoted prediction that the twenty-first century would be an era of water wars might seem to have proven unfounded. War continues to plague the planet, however, in the Middle East, Afghanistan, the Ukraine and elsewhere. In many cases, if not most, these are wars for control of natural resources, though they are usually represented as having quite different causes.

Israel's war on its neighbours were about the water that was indispensable to its desert economic project. The 1967 conflict over the Golan Heights had as its unacknowledged object the control of the source of the River Jordan. Today its domination of the waters of the Jordan has become an instrument in another, ongoing war – against the Palestinian people, trapped by thirst and an unrelenting military machine. In the Sudan, the assaults by the government of Sudan on the south were motivated by the discovery of oil, but thirst was a powerful weapon in its hands. The coming battle for the Arctic and Antarctic will also be resource wars, and directly and indirectly a battle for water too. And the predominant arrangements for water distribution along the major rivers, now dominated by ideas of 'water capture' – that water belongs to the landowner to be sold at will as a commodity – will certainly generate conflicts as pressure grows. There is also a sort of neo-Malthusian nightmare of war that results from population growth. Hollywood has produced dozens of films that act out that fantasy, showing cities under siege by zombies who bear a striking resemblance to the poor. All of them are reworkings of the fear of the 'barbarians at the gates'. Talk of water wars often seems to reproduce the night terrors of the western world.

The alternative is what we might call 'water peace'. To take the case of rivers, there are very different principles available for the fair allocation of their water. The 'riparian' principle and the 'acequia' system are based on social justice, and an equal distribution of water. It is of course absurd to suggest that a water-table, which stretches for hundreds of kilometres in every direction, can belong to the owner of one piece of land. It is one more reason why water cannot be owned, and thus cannot be bought and sold or privatised in any sense.

Neo-liberalism, global capitalism, is driven by two motor forces – accumulation and competition, the polar opposites of cooperation and distribution according to need. Yet a water order can only be based on these latter principles. We are told that privatisation will conserve water through a price mechanism. It is a curious use of the word conserve, which in this case would be better replaced by hoard; if water is scarce, then let the rich have it and the poor go without. The World Bank, the IMF and the other international financial institutions clearly work on that basis, providing funds only where there is a guarantee of a return on it in what are inaptly named Public–Private Partnerships. The alternative, PUPs, or Public–Public Partnerships are built on non-profit exchanges and collaborations between public bodies. But these can only function properly where water is unequivocally a public good, and cannot be privatised.

There is a persistent and growing body of opinion that argues that price alone can ensure a proper water supply. We have argued that it has two foundations: one, that water can be saved by depriving the poor of it; and two, that public utilities are inherently unable to develop proper mechanisms of supply and distribution.[4] Yet they managed to do so in much of the world for most of the last 2,000 years. It is an ideological position, veiled as objective fact. Many municipal providers work very well, others less so; the record of private corporations is, if anything, significantly worse. And the method of allocation by agreement seems to have worked pretty well in Valencia, Spain, where the water court first established by the Moors has met on the cathedral steps every Thursday for a thousand years!

The repair of urban water infrastructure could resolve the water supply of cities, repairing pipes and stopping leaks as soon as they appear. But that is a public responsibility. As Tokyo and Singapore have found, it could make a significant difference.

Local farm practices are not primitive leftovers but systems of cultivation that have fed local populations for centuries. Rainwater harvesting in India and the capture of dew and mist in Chile have been extremely successful, as have alternative farming methods using drip irrigation and no till farming. They should be explored, investigated and respected and included in a repertoire of methods appropriate to their time and place. What is interesting is that these practices are often described in near-racist ways as remnants of folk culture. It is curious then that in the High Andes the terraces the Incas built have been restored because they were found to be the most efficient cultivation method for that landscape, or that the colonial system of acequias have been recently researched and found to be not just environmentally friendly but also extraordinarily efficient in conserving water. These communities of farmers will often combine many different forms of operation and many different methods of negotiation with local and national state, skilfully managing their water. The universal methods imposed by the market, by contrast, often fail in the medium to long term and are invariably costly – as a glance at the current state of the Amazon will show.

And further ahead …

In the end, capitalism is the enemy of social justice, of democracy, and the scourge of the planet. As the struggle against waste, against the poisoning of the earth, for the protection of water develops it will become increasingly clear that capitalism and its relentless pursuit of profit stands in the way of a balanced and harmonious environment. The struggle itself, the increasing involvement of people across the world for democracy and local control, will build the democratic structures on which a new world can be founded.

As the respected South African water campaigner Trevor Ngwane, of the Anti-Privatisation Forum, puts it:

> I [have] tried to avoid preaching to the converted; my imagined target audience is 'a fresh mind', say, a young person still learning the ABC of the struggle, or an older comrade who has been too busy struggling on the ground to give much thought to these issues. To such comrades I say: the time has come to take up the struggle to save the earth and to safeguard nature from capitalist destruction and its structured ignorance. Animals and plants are part of nature. Human beings are also part of nature, they too inhabit the earth. We need a vision of a world where

humans, animals, plants, forests, rivers, mountains, valleys and all other aspects of nature live harmoniously together. We cannot turn the clock back to the idyllic and uncomplicated stage of primitive communism. But we can embrace the idea of eco-socialism and struggle to realize it practically in order to advance to communism – the classless society.[5]

Notes

Introduction

1. These are the often quoted figures from the Millennium Development Goals documents, see: www.un.org/millenniumgoals (accessed 15 March 2015).
2. Maggie Black, *No-Nonsense Guide to Water* (London: Verso, 2004).
3. Alexander Bell, *Peak Water: Civilisation and the World's Water Crisis* (Edinburgh: Luath, 2009).
4. Oscar Olivera with Tom Lewis, *Cochabamba! Water War in Bolivia* (New York: South End Press, 2004).
5. These words were delivered in 1688 by Winstanley, the leader of the Diggers, a utopian religious sect.
6. Asit K. Biswas found something of the order of 150 million texts on water available on the internet! See his lecture on 'The Future of the World's Water: Rhetoric and Reality?', 2 May 2013, The Water Institute, at: www.water.uwaterloo.ca
7. Constance Elizabeth Hunt, *Thirsty Planet: Strategies for Sustainable Water Development* (London: Zed Books, 2004).
8. Hunt (2004), p. 40.
9. A product can have a 'use value', which may satisfy human needs or 'be a means of subsistence'. Its 'exchange value' is determined in the market, and goes beyond the cost of its production. It becomes a 'commodity' at that point, its value expressed in price not usefulness. The difference between the cost of producing it and the sale price is the 'profit' (or surplus value as Marx called it) around whose accumulation capitalism as a system is organised.

1. A floating planet

1. Peter Gleick, *et al.*, *The World's Water 2014: The Biennial Report on Freshwater Resources*, vol. 8 (Washington, DC: Island Press, 2014), p. 3.
2. There is an obvious problem trying to imagine a cubic metre or a cubic kilometre of water. Fred Pearce suggests we think of three baths full of water as 1 cubic metre (it takes 3,125 to fill an Olympic swimming pool). A cubic kilometre is even more inaccessible; the Nile contains 50.
3. These figures are taken from Maggie Black's *No-Nonsense Guide to Water* (London: Verso, 2004), which tends to summarise and present statistics in the most accessible way. See also, Fred Pearce, *When the Rivers Run Dry:*

What Happens When our Water Runs Out? (London: Eden Project Books, 2007).

4. According to J. A. 'Tony' Allan, *Virtual Water: Tackling the Threat to Our Planet's Most Precious Resource* (New York/London: IB Tauris, 2011).
5. Black, *No-Nonsense Guide to Water*, p. 19.
6. Alexander Bell, *Peak Water: Civilisation and the World's Water Crisis* (Edinburgh: Luath, 2009).
7. See Kaye LaFond, 'High Levels of Benzene in California Fracking Waste Water', 12 February 2015, at: www.circleofblue.org (accessed 28 February 2015).
8. See Kaye LaFond, 'Humans Degrading the Systems that Sustain Them', 16 January 2015, at: www.circleofblue.org (accessed 24 January 15); Black, *No-Nonsense Guide to Water*, p. 23.
9. Pearce, *When the Rivers Run Dry*, Chapter 3. The typical Western consumer he mentions is himself!
10. See Black, *No-Nonsense Guide to Water*.
11. Black, *No-Nonsense Guide to Water*, p. 20.
12. It would be interesting to know whether their location might explain the scant attention that has been paid to this environmental tragedy!
13. See Pearce, *When the Rivers Run Dry*, Chapter 12; also Kaye LaFond, 'Florida State Employees Can't Say "Climate Change"', 12 March 2015, at: www.circleofblue.org (accessed 15 April 2015).
14. Black, *No-Nonsense Guide to Water*, p. 32.
15. Steve Maxwell with Scott Yates, *The Future of Water: A Startling Look Ahead* (Denver, CO: American Water Works Association, 2011).
16. See Chapter 7.
17. We return to this in Chapter 5.
18. Asit K. Biswas, 'The Future of the World's Water: Rhetoric and Reality?', 2 May 2013, The Water Institute, at: www.water.uwaterloo.ca
19. The term often used is 'individuation'.
20. David Harvey, *The New Imperialism* (Oxford: Oxford University Press, 2003), p. 148.
21. David Harvey, *A Brief History of Neoliberalism* (Oxford: Oxford University Press, 2007).

2. How water was privatised

1. James Salzman, *Drinking Water: A History* (London/New York: Duckworth Overlook, 2012).
2. Ann-Christin Sjölander Holland, *The Water Business: Corporations Versus People* (London/New York: Zed Books, 2005) estimates that in the USA one quarter of bottled water comes directly from the tap.

3. Steve Maxwell with Scott Yates *The Future of Water: A Startling Look Ahead* (Denver, CO: American Water Works Association, 2011), p. 45.

4. Anya Groner, 'The Politics of Drinking Water', *The Atlantic*, 30 December 2014. At this point (2015) petrol in the USA costs around $3 a gallon.

5. Daniel Azpiazu (ed.), *Las privatizaciones en la Argentina: Diagnóstico y propuestas para una mayor competitividad y equidad social* (Buenos Aires: Fundación OSDE, 2002).

6. Sjölander Holland, *The Water Business*, pp. 11–12.

7. David Hall and Emanuele Lobina, *Water Privatisation in Latin America*. Public Services Research Institute, University of Greenwich, London, July 2002. p. 10. This study is specific to Latin America but their conclusions are applicable across the world.

8. 'In many cases, privatisation is undertaken in response to conditions set up by international institutions, such as the World Bank, the IMF and other development banks.' Sjölander Holland, *The Water Business*, p. 83.

9. This was also the rationalisation given in earlier decades by the World Bank to foster the building of dams, but this is an issue we will return to.

10. Hall and Lobina cite the case of Nicaragua, where the government officials responsible for water formed their own private companies and bid successfully for contracts in two Nicaraguan towns. As the Trans-Nicaragua Canal project nears the moment of its inauguration, at a cost of $40 billion, there is likely to be further crossover of officials, as well as a serious drinking water crisis as the canal makes its way across Lake Nicaragua, the main source of the country's drinking water.

11. Sjölander Holland, *The Water Business*, p. 117.

12. Hall and Lobina, *Water Privatisation in Latin America*, p. 10.

13. Ibid., p. 4.

14. John Vidal, 'Water Privatisation: A Worldwide Failure?', *Guardian*, 30 January 2015.

15. See Satoko Kishimoto, 'World Water Forum Needs to be More than Just a Trade Show for Privatisation', *Guardian*, 17 April 2015.

16. Maude Barlow and Tony Clarke, *Blue Gold: The Battle Against Corporate Theft of the World's Water* (London/New York: The New Press, 2003), p. 69.

17. Quoted in John Vidal, 'Water Privatization: A Worldwide Failure?', *Guardian*, 30 January 2015.

18. Vidal, 'Water Privatisation'.

19. According to the latest estimates of the WHO/UNICEF Joint Monitoring Programme for Water Supply and Sanitation (JMP), released in early 2013 (collected in 2011), at: www.unicef.org/wash/ (accessed 15 March 2015).

20. Kurt Hollander, 'Mexico City: Water Torture on a Grand and Ludicrous Scale', *Guardian*, 5 February 2014.

21. See Juan Pablo Orrego S., 'La crisis mundial de las aguas: algunas de sus principales causas', Chile, March 2013, for the Fundación Terram, at: www. terram.cl (accessed 16 February 2015).
22. Francis Fukuyama, *The End of History and the Last Man* (London/New York: Free Press, 1992).
23. See Taryn Luntz, 'US Drinking Water Widely Contaminated', *Scientific American*, 14 December 2009, at: www.scientificamerican.com (accessed 1 May 2015).
24. Scott Smith, 'California Shuts Down Oil Wells to Protect Groundwater', *Huffington Post*, 3 March 2015, at: www.huffingtonpost.com
25. Anthony Giddens, *The Politics of Climate Change* (Cambridge: Polity Press, 2011; 2nd edn) p. 9.

3. Disasters, natural and otherwise

1. Helena Lindholm, 'Water and the Arab–Israeli conflict', in Leif Ohlsson (ed.), *Hydropolitics: Conflicts over Water as a Development Constraint* (London: Zed Books, 1995), p. 61. See also, 'Down the Drain', a report on Israeli restrictions on the water and sanitation (EWASH) sector in the occupied territories, March 2012, at: www.ewash.org (accessed 28 February 2015); Diane Raines Ward, *Water Wars: Drought, Flood, Folly, and the Politics of Thirst* (New York: Riverhead Books, 2002), pp. 187–88.
2. See 'Gaza facts & figures', UNICEF oPt, November 2012, at: www.unicef.org/oPt/UNICEF_oPt_-_Gaza_Fact_sheet_-_November_2012.pdf (accessed 15 March 2015).
3. 'Water Crisis: Discriminatory Water Supply', *B'Tselem* – The Israeli Information Center for Human Rights in the Occupied Territories, 10 March 2014, at: www.btselem.org/water/discrimination_in_water_supply (accessed 31 May 2015).
4. According to World Watch, a general improvement of 10 per cent in the use of water for irrigation would satisfy the personal water needs of everyone on the planet. Raines Ward, *Water Wars*, p. 121.
5. Rachel Carson, *Silent Spring* (Boston, MA: Houghton Mifflin, 1962).
6. This is what Anthony Giddens calls, a little arrogantly, *Giddens's paradox* – '[when] the dangers posed aren't tangible, immediate or visible in the course of day to day life, many will sit on their hands and do nothing of a concrete nature about them.' Anthony Giddens, *The Politics of Climate Change* (Cambridge: Polity Press, 2011; 2nd edn), p. 2.
7. Constance Elizabeth Hunt, *Thirsty Planet: Strategies for Sustainable Water Development* (London: Zed Books, 2004), p. 1.
8. Hunt, *Thirsty Planet*, p. 2.
9. Basil Davidson, *Africa in History* (New York: Touchstone, 1995).

10. Hunt, *Thirsty Planet*, p. 38, quoting Fred Pearce's *The Dammed: Rivers, Dams and the Coming World Water Crisis* (London: Bodley Head, 1992).

11. See Eduardo Galeano, *The Open Veins of Latin America: Five Centuries of the Pillage of a Continent* (New York: Monthly Review Press, 1973) republished in 2008 (London: Latin America Bureau).

12. See Chapter 6.

13. The emblematic description of that early ecological disaster and its human consequences was John Steinbeck's *The Grapes of Wrath* and echoed in Woody Guthrie's poignant *Dustbowl Ballads*.

14. John Wesley Powell (1834–1902) was a geologist and anthropologist who charted the course of the Colorado River in 1869 and gave the Grand Canyon its name. As a conservationist *avant la lettre* he argued that the arid regions of the west would not support farming and warned that 'you are piling up a heritage of conflict and litigation over water rights, for there is not sufficient water to supply the land.' Powell did propose large-scale irrigation, but based on watersheds not state boundaries. The railway companies, with far greater influence over Congress, won the argument, however. (See Michael Hiltzik, 'The False Promise of Hoover Dam', *Los Angeles Times*, 5 July 2010). His ideas about native Americans, however, were anything but progressive, though he was appointed Professor of Ethnography at the Smithsonian.

15. The term, derived from the Arabic, described the method for allocating water established during the Moorish occupation of southern Spain, and which survives in the Valencia water tribunal, see Chapter 10.

16. Devon G. Peña, 'Revolutions Happen: On the Crisis of Neoliberalism and the Alternative of the Common', at: www.newclearvision.com/2014/04/25/revolutions-happen (accessed 29 April 2015).

17. It is the prevailing principle in Texas, for example, where the recent prolonged drought called it briefly into question until the return of the rains (in 2015) restored the dominion of private ownership and water capture!

18. Steven Solomon, *Water: The Epic Struggle for Wealth, Power, and Civilization* (New York: HarperCollins, 2010), pp. 342–44.

19. It was Nehru, the first president of independent India, who used this famous analogy, though he came to regret it later.

20. In 1973 Richard Nixon signed the Endangered Species Act and under Ronald Reagan a limit was put on federal spending for dams, raising the contribution of individual states. So there were both financial and environmental issues at play in the decision. Since then, 1125 dams have been dismantled in the USA.

21. Arundhati Roy, 'The Greater Common Good', in *The Cost of Living* (London: Flamingo, 1999).

22. Fred Pearce, *When the Rivers Run Dry: What Happens When our Water Runs Out?* (London: Eden Project Books, 2007).

23. See Asit K. Biswas, 'Impacts of Large Dams: Issues, Opportunities and Constraints', in Cecilia Tortajada *et al.* (eds), *Impacts of Large Dams: A Global Assessment* (Berlin: Springer, 2012), pp. 1–18.

24. Ibid., p. 7.

25. Pearce, *When the Rivers Run Dry*.

26. Raines Ward, *Water Wars*, p. 81.

27. And in 2014 the main reservoir supplying São Paulo was at 10 per cent of its initial volume, forcing water rationing on the city's 20 million inhabitants. See Chapter 4 on the Amazon.

28. This figure is at the lower end of available estimates. See Robin Clarke and Jannet King, *The Atlas of Water: Mapping the World's Most Critical Resource* (London: Earthscan, 2004).

29. J. A. Allan sees it as a direct result of the one-child policy, hence his assertion that, in terms of water, China is 'saving the world', a view we find less than convincing. J. A. 'Tony' Allan, *Virtual Water: Tackling the Threat to Our Planet's Most Precious Resource* (New York/London: IB Tauris, 2011).

30. See Chapter 4.

31. See, for example, Kevin P. Gallagher and Roberto Porzecanski, *The Dragon in the Room: China & the Future of Latin American Industrialization* (Stanford, CA: Stanford University Press, 2010).

32. Charlton Lewis, 'China's Great Dam Boom: A Major Assault on its Rivers', *Yale Environment 360*, 4 November 2013, at: www.e360.yale.edu (accessed 15 March 2015).

33. Pearce, *When the Rivers Run Dry*.

34. Kenneth Pomeranz, 'The Great Himalayan Watershed: Agrarian Crisis, Mega-Dams and the Environment', *New Left Review*, 58, July–August 2009, pp. 5–39.

35. See Leslie Hook, 'China: High and Dry', *Financial Times*, 14 May 2013.

36. Pearce, *When the Rivers Run Dry*, Chapter 14.

37. Peter H. Gleick, *The World's Water: The Biennial Report on Freshwater Resources*, vol. 7, (Washington, DC: Island Press, 2011), Chapter 5, 'China and Water'.

38. As compared to 11 per 100,000 in Vietnam and 5 per 100,000 in Thailand.

39. The very recent corruption scandal in Brazil's state-owned oil company Petrobras, exposed corruption on a similar scale in the construction of the giant Belo Monte Dam in the Amazon.

40. Interview with Vandana Shiva, at: www.inmotionmagazine.com (accessed 15 March 2015). See too her important contribution to the current debate in, Vandana Shiva, *Water Wars: Privatization, Pollution, and Profit* (London: Pluto Press, 2002).

41. This is a position echoed by Ricardo Petrella who refers to the 'new lords of water' for whom water is a source of power, wealth and domination. Ricardo Petrella, 'Globalization and Internationalization: The Dynamics of

the Emerging World Order', in Robert Boyer and Daniel Drache (eds), *States Against Markets: The Limits of Globalization* (London, Routledge: 1996), pp. 62–83.

42. Roy, 'The Greater Common Good'.

43. Clarke and King, *The Atlas of Water*, p. 87.

44. Vandana Shiva, 'The New Food Wars: Globalization GMOs and Biofuels', video, *GM Watch*, at: www.gmwatch.org (accessed 15 March 2015).

45. Palagummi Sainath, 'A 12-year Saga of Farm Suicides in India', The Institute of Science in Society, ISIS Press Release, 9 February 2010, at: www.i-sis.org.uk/Farm_Suicides_in_india.php (accessed 30 March 2015).

46. Pearce, *When the Rivers Run Dry*.

47. Pomeranz, 'The Great Himalayan Watershed'.

48. See Ricardo Petrella, 'Globalization and Internationalization: The Dynamics of the Emerging World Order', pp. 62–83.

49. Black, *No-Nonsense Guide to Water*, p. 44; Pearce, *When the Rivers Run Dry*.

50. Pearce, *When the Rivers Run Dry*.

51. Steve Maxwell with Scott Yates, *The Future of Water: A Startling Look Ahead* (Denver, CO: American Water Works Association, 2011), p. 117.

52. Hunt, *Thirsty Planet*, p. 98.

53. Maxwell and Yates, *The Future of Water*, pp. 186–87.

54. Ibid., p. 187.

55. Black, *No-Nonsense Guide to Water*, pp. 34–36.

56. A friend describes the perplexity of an NGO providing a village well in Africa; the local women had previously walked ten kilometres each way to collect water, but the new well was regularly sabotaged. It emerged that the women themselves had filled the new well with stones because their only opportunity to meet without their husbands was walking to the well.

57. Blacksmith Institute and Green Cross International, at: www.worldsworstpolluted.org (accessed 9 March 2015).

58. Ibid.

4. A short trip through Amazonia

1. Some examples might be the films *At Play in the Fields of the Lord* (Director: Hector Barbenco, 1991) and *Medicine Man* (Director: John McTiernan, 1992).

2. Henry Ford built an exemplary small American town, Fordlandia, on the Amazon in the 1930s in the hope of developing a rubber industry there. The fascinating story of Fordlandia's rise and fall is told in Greg Grandin's *Fordlandia: The Rise and Fall of Henry Ford's Forgotten Jungle City* (New York: Metropolitan Books, 2009).

3. Olimar E. Maisonet-Guzman, 'Amazon Battle: Is Hydropower the New Kobayashi Maru?' *E-International Relations*, 17 December 2010, at: www.e-1r.info (accessed 25 February 2015).

4. Philip M. Fearnside, Adriano M. R. Figueiredo and Sandra C. M. Bonjour, 'Amazonian Forest Loss and the Long Reach of China's Influence', *Environment, Development and Sustainability*, 15 (2013), pp. 325–38, at: http://static.springer.com/sgw/documents/1380546/application/pdf/ Amazonian.pdf (accessed 18 March 2015).

5. Simón Farabundo Ríos, 'Integration and the Environment on the Rio Madeira', *NACLA*, 9 March 2009, at: https://nacla.org/node/5595 (accessed 15 March 2015).

6. Eduardo Galeano, *The Open Veins of Latin America: Five Centuries of the Pillage of a Continent* (New York: Monthly Review Press, 1973) republished in 2008 (Latin America Bureau, London).

7. Edward Docx, 'The Last Stand of the Amazon', *Guardian*, 3 April 2011.

8. And there are good reasons to suspect that large swathes of land, previously used to cultivate crops like potatoes, are now given over to poppy growing.

9. It appears to have been unaffected by the Ice Age, hence the unique range of flora and fauna still to be found there.

10. See Marcela Olivera's 'Water beyond the State: Letter from Cochabamba', pp. 104–8.

11. See Boletim na falta d'Agua em SP by @camilalpav, at: http://boletim dafaltadagua.tumblr.com/

12. See Yuri Leveratto, 'La Amazonía colombiana, nueva frontera de la explotación minera y petrolera', 2012, at: http://yurileveratto.com/articolo. php?Id=290/ (accessed 15 March 2015).

13. See Rudolf von May and Rhett Butler, 'Deforestación en aumento en mayoría de países amazónicos', Mongabay.com, 15 August 2013, at: http:// es.mongabay.com/news/2013/0814-deforestacion-en-paises-amazonicos. html (accessed 15 March 2015).

5. *Bitter harvests*

1. Amartya Sen, the Nobel prize-winning economist, has reiterated and developed similar conclusions. See his *Poverty and Famines: An Essay on Entitlement and Deprivation* (Oxford: Oxford University Press, 1981).

2. For example, in Garrett Hardin's widely disseminated essay, 'The Tragedy of the Commons', published in the magazine *Science*, 162: 3859 (13 December 1968).

3. See Mark Dowie, *American Foundations: An Investigative History*, (Boston, MA: MIT Press, 2001); and John H. Perkins, *Geopolitics and the Green Revolution: Wheat, Genes, and the Cold War* (Oxford: Oxford University Press, 1997).

4. 2013 Report of Blacksmith Institute, New York, at: www.worstpolluted.org
5. Vandana Shiva, 'The New Food Wars: Globalization GMOs and Biofuels', video, *GM Watch*, at: www.gmwatch.org (accessed 15 March 2015).
6. Who produced the infamous napalm and Agent Orange, whose impact affected not only human beings but also devastated the soil of Vietnam for more than a generation.
7. Vandana Shiva on climate change, 'Time to End the War against the Earth', *The Age*, 4 November 2010, at: www.theage.com.au
8. World Resources Institute Report 2013–2015: Creating a Sustainable Food Future, at: www.wri.org (accessed 2 March 2015).
9. Tim Searchinger and Ralph Heimlich, 'Avoiding Bioenergy Competition for Food Crops and Land', *Creating a Sustainable Food Future*, Installment 9, World Resources Institute 2015, at: www.wri.org (accessed 2 March 2015).
10. Constance Elizabeth Hunt, *Thirsty Planet: Strategies for Sustainable Water Development* (London: Zed Books, 2004), Chapter 3; Peter Gleick, *et al.*, *The World's Water 2014: The Biennial Report on Freshwater Resources*, vol. 8 (Washington, DC: Island Press, 2014), at: www.worldwater.org
11. The uprising of the Zapatista communities in Chiapas, Mexico in 1994 was directly the result of the dramatic impoverishment of peasant farmers. See Tom Hayden (ed.), *The Zapatista Reader* (New York: Thunder's Mouth Press/ Nation Books, 2002).
12. Hunt, *Thirsty Planet*, p. 65.
13. See, for example, *World Resources Report 2013–15*, p. 6.
14. The outstanding work of Sandra Postel has addressed in the greatest detail the shortcomings and deleterious effects of mismanaged irrigation. See for example, her 'Water and agriculture' in P. H. Gleick (ed.), *Water in Crisis: A Guide to the World's Freshwater Resources* (New York: Oxford University Press, 1993).
15. As Myrdal puts it in his 1975 Nobel Prize acceptance speech, 'The technocratic euphoria some ten years ago about a "green revolution" had already earlier been shown up as having nurtured undue optimism.' See www.nobelprize.org
16. Hunt, *Thirsty Planet*, p. 70.
17. Ibid.
18. Quoted in Vandana Shiva's lecture 'Earth Democracy' at Portland Community College, Portland, Oregon, 24 February 2011, at: www.youtube.com/watch?v=UOfM7QD7-kk (accessed 15 March 2015).
19. Vandana Shiva, 'New Food Wars: Globalization GMOS and Biofuels', University of California at Irvine lecture on 3 July 2008, University of California Television (UCTV), at: www.youtube.com/watch?v=Iq6jpkDNxtI (accessed 15 March 2015).
20. Julian Caldecott, *Water: Life in Every Drop* (London: Virgin Books, 2007).

21. *World Water 2013, vol. 8, 2013*, Table 2, 'Freshwater Withdrawal by Country and Sector', at: www.worldwater.org (accessed 28 March 2015).

22. Figures from IPCC, *Climate Change 2007: mitigation*, quoted in Jonathan Neale, *Stop Global Warming: Change the World* (London: Bookmarks, 2008), p. 92.

23. Black, *No-Nonsense Guide*, p. 100.

24. World Water Assessment Programme, at: www.unesco.org/water/wwap

25. Frank Piasecki Poulsen's film *Blood in the Mobile* (Denmark, 2010) deals with this terrible trade. See *Coltan el mineral de la muerte*, at: http://simiomobile.com/

26. See *La Revista Minera*, 2009–2013, at: https://revistaminera.wordpress.com

27. Safe Drinking Water Foundation, Canada, at: www.safewater.org (accessed 17 March 2015).

28. Ibid.

29. Systematically exposed by Mining Watch, Canada, from which much of the material quoted here is taken, at: www.miningwatch.ca

30. The Cuenca resistance was part of the struggle of the Shuar nation against the invasion of its territories and the contamination of its water by foreign mining and oil companies. It is ironic that their activities were criminalised, and their activists jailed, by the supposedly progressive government headed by Rafael Correa.

31. See Earthworks: No Dirty Gold, at: http://nodirtygold.earthworksaction.org

32. Neale, *Stop Global Warming*, p. 153.

33. There is an interesting discussion on fracking sponsored by the US Geological survey, Melanie Gade (Video Producer), 'Science or Soundbite? Shale Gas, Hydraulic Fracturing, and Induced Earthquakes', 4 April 2012, at: http://gallery.usgs.gov/videos/533#.VQjlh0uQbwI

6. Virtual water

1. Arjen Y. Hoekstra, Ashok K. Chapagain, Maite M. Aldaya and Mesfin M. Mekonnen, *The Water Footprint Assessment Manual: Setting the Global Standard* (London: Earthscan, 2011), at: http://waterfootprint.org/

2. Ibid.

3. See Chapter 2.

4. J. A. 'Tony' Allan, *Virtual Water: Tackling the Threat to Our Planet's Most Precious Resource* (New York/London: IB Tauris, 2011).

5. And Pearce and others use it too.

6. Morgan Spurlock's film *Super Size Me* (2004) is a powerful counter to McDonald's publicity.

7. See Chapter 4.

8. Saudi Arabia's attempts to grow wheat and develop a beef industry proved immensely costly and inefficient and have been abandoned. It has returned to buying all its needs with oil revenues.

9. As we point out in Chapter 5, Israel also exports its water scarcity to the Palestinians, whose thirst has watered the desert on the other side of the Wall.

10. Julian Caldecott, *Water: Life in Every Drop* (London: Virgin Books, 2007).

11. Ibid., pp. 153–54.

12. See Vandana Shiva, *The Violence of the Green Revolution: Third World Agriculture, Ecology and Politics* (London: Zed Books, 1991).

13. According to Professor Asit Biswas the highest individual consumption in the world is, astonishingly, in Qatar where average consumption tops 400 litres! Asit K. Biswas, 'Future of the World's Water: Rhetoric and Reality?', 2 May 2013, The Water Institute, at: www.water.uwaterloo.ca

14. We will return in Chapter 8 to a longer discussion of 'water wars' and their meaning.

15. The concept of 'comparative advantage' encapsulates how economic decisions are made under capitalism – it refers to competition, and the advantage one capitalist has over another in terms of costs of production.

16. 'The Concept of Virtual Water: A Critical Review', January 2008, Department of Environment, Land, Water & Planning, at: www.depi.vic.gov.au (accessed 18 March 2015).

17. 'Conclusions' in 'The Concept of Virtual Water: A Critical Review', January 2008, Department of Environment, Land, Water & Planning, at: www.depi. vic.gov.au

18. Lena Horlemann and Susanne Neubert, *Virtual Water: A Realistic Concept for Resolving the Water Crisis?*, German Development Institute, 2007, at: edoc.vifapol.de/opus/volltexte/2013/4368/ (accessed 15 March 2015).

19. For more on the role of the WTO, see below.

20. There is of course a growing trade in *actual* water, which we have addressed in Chapter 2.

21. See Rutgerd Boelens, *et al.*, 'Introduction', in Rutgerd Boelens, David Getches and Armando Guevara-Gil (eds), *Out of the Mainstream: Water Rights, Politics and Identity* (London: Earthscan, 2010), p. 10.

22. 'Accumulate, accumulate, that is Moses and the Prophets', in Karl Marx, *Capital vol. 1.*

7. Water and global warming

1. Jonathan Neale, *Stop Global Warming: Change the World* (London: Bookmarks, 2008), p. 164.

2. See Fred Pearce, *The Climate Files: The Battle for the Truth about Global Warming* (London: Guardian Books, 2010). The book explores what came to be called 'Climategate' when a decade and a half of emails between climate scientists in the UK and the USA were leaked to the press, and fuelled a renewed campaign of climate change denial. His conclusion, after a painstaking investigation of 1,073 emails and 3,587 document files, is that 'nothing uncovered in the emails destroys the argument that humans are warming the planet'.

3. Pearce, *The Climate Files*, Chapter 19.

4. Published in Denmark's newspaper, *Dagbladet Information*.

5. Fred Pearce, *When the Rivers Run Dry: What Happens When our Water Runs Out?* (London: Eden Project Books, 2007), Chapter 15.

6. These figures are from Pearce, *When the Rivers Run Dry*, Chapter 15.

7. Kenneth Pomeranz, 'The Great Himalayan Watershed: Agrarian Crisis, Mega-Dams and the Environment', *New Left Review*, 58 (July/August 2009).

8. Shilong Piao, *et al.*, 'The Impacts of Climate Change on Water Resources and Agriculture in China', *Nature*, 467: 2 (September 2010), pp. 43–51.

9. Alun Anderson, *After the Ice: Life, Death and Politics in the New Arctic* (London: Virgin Books, 2009), p. 75.

10. See the official tourist site, at: www.norilsk.net and the Blacksmith Institute assessment, at: www.blacksmithinstitute.org

11. Philip M. Fearnside, Adriano M. R. Figueiredo and Sandra C. M. Bonjour, 'Amazonian Forest Loss and the Long Reach of China's Influnce', *Environment, Development and Sustainability*, 15 (2013) pp. 325–38, at: http://static.springer.com/sgw/documents/1380546/application/pdf/Amazonian.pdf (accessed 18 March 2015).

12. Vandana Shiva, 'Soil Not Oil: Climate Change, Peak Oil and Food Justice', on 23 February 2011. World Affairs Council of Oregon, at: www.worldoregon.org (accessed 18 March 2015).

13. Jonathan Neale describes how the American soft drinks and brewing industries developed disposables (plastic bottles and tin cans) to replace the glass bottles which were returnable, washed and then re-used. 'The cost of packaging rose to $26 billion a year, a substantial part of the output of American industry. All this plastic packaging, and all the bottles and cans, counted as economic growth. But they were waste of no use to anyone.' Neale, *Stop Global Warming*, p. 206. Their legacy are the mountains, cliffs and acres of garbage that now 'litter' the world – the word itself an invention of that same industry.

14. Shilong Piao, *et al.*, 'The Impacts of Climate Change on Water Resources and Agriculture in China'.

15. Michal Kravcik is interviewed at length in the film *Blue Gold* (2004), based on Maude Barlow and Tony Clarke's book of the same name. See his talk

'Water and the Recovery of the Planet' at the University of Sydney's Carbon and Water Symposiumm May 2011, at: www.youtube.com

16. Juan Pablo Orrego, 'La crisis mundial de las aguas: algunas de sus principales causas', Chile, March 2013, for the Fundación Terram at: www.terram.cl (accessed 16 February 2015).

17. See Asit K. Biswas, 'Water for a Thirsty Urban World', *Brown Journal of World Affairs*, 17: 1 (Fall–Winter 2010).

18. Ibid.

19. In Biswas' lecture 'Climate Change and Water Management', on 16 October 2009, at: www.thirdworldcentre.org

20. Ibid.

21. Quoted in 'Transitions Towards Adaptive Management of Water Facing Climate and Global Change', at: www.indiaenvironmentportal.org.in. Gleick's original formulation is in 'Global Freshwater Resources: Soft-Path Solutions for the 21st Century', *Science*, 302 (28 November 2003), pp. 524–28.

8. Ya basta! Enough is enough!

1. Oscar Olivera with Tom Lewis, *Cochabamba! Water War in Bolivia* (New York: South End Press, 2004).

2. For more on the Zapatistas, see Tom Hayden (ed.), *The Zapatista Reader* (New York: Thunder's Mouth Press/Nation Books, 2002) and Subcomandante Marcos, *Our Word is Our Weapon: Selected Writings*, edited by Juana Ponce de León (New York: Seven Stories Press, 2004), a collection of the writings of the charismatic Marcos, the acknowledged spokesperson (but not the leader) of the movement.

3. It could equally be argued that the first rebellion against globalisation was the *Caracazo* in Venezuela, a general insurrection of the poor against economic measures imposed by the International Monetary Fund. It was repressed within three days, leaving a death toll in the thousands.

4. Jim Shultz, 'The Politics of Water in Bolivia: Once Again, World Bank Water Policy is Challenged by the Poorest', *The Nation*, 28 January 2005. Schulz, a North American, lives in Cochabamba and has sustained his informative website, The Democracy Center, for many years, at: www.democracyctr.org

5. This was probably true, but there is an issue here of who should cast the first stone!

6. Hence the title of a wonderful film about Cochabamba, *También la lluvia* … (Even the rain …), (Director: Icíar Bollaín, 2010).

7. Jean Friedman-Rudovsky, 'Return to Cochabamba: Eight Years Later, the Bolivian Water War Continues', *Earth Island Journal*, The Water Issue, 23: 3 (Autumn 2008), at: www.earthisland.org/journal/index.php/issues/toc/autumn_2008/

8. See Rutgerd Boelens, David Getches and Armando Guevara-Gil (eds), *Out of the Mainstream: Water Rights, Politics and Identity* (London: Earthscan, 2010).
9. Justin Doolittle, 'The Corporate Assault on Latin American Democracy', *counterpunch*, 3 November 2014, at: www.counterpunch.org/2014/11/03/the-corporate-assault-on-latin-american-democracy/ (accessed 15 March 2015).
10. José Esteban Castro, 'Water Struggles, Citizenship and Governance in Latin America', *Development*, 51 (March 2008), pp. 72–76.
11. Asit K. Biswas, 'The Future of the World's Water: Rhetoric and Reality?', 2 May 2013, The Water Institute, at: www.water.uwaterloo.ca
12. Ibid.
13. See Julia Apland Hitz, 'The Water Conflict in Ecuador', *State of the Planet*, 14 May 2010, at: http://blogs.ei.columbia.edu/2010/05/14/the-water-conflict-in-ecuador/ (accessed 10 April 2015).
14. Alberto Acosta, 'El agua, un derecho humano no un negocio', 12 May 2010, *Rebelión*, at: www.rebelion.org/noticias/2010/5/105741.pdf (accessed 20 March 2015).
15. Willem Assies, 'The Limits of State Reform and Multiculturalism in Latin America: Contemporary Illustrations', in Rutgerd Boelens, *et al.*, *Out of the Mainstream: Water Rights, Politics and Identity* (London: Earthscan, 2010), p. 61.
16. See Marcela Olivera's 'Water against the State: Letter from Cochabamba', which addresses these issues, pp. 104–8.
17. Discussed in Chapter 4.
18. Anthony Bebbington, Denise Humphreys Bebbington and Jeffrey Bury, 'Federating and Defending: Water, Territory and Extraction in the Andes', in Rutgerd Boelens, *et al.*, *Out of the Mainstream*, p. 313.
19. For this section we have relied heavily on Marcelle Dawson's excellent work on the Anti-Privatisation Forum, which she was kind enough to share with us. See for example, 'The Cost of Belonging: Exploring Class and Citizenship in Soweto's Water War' in *Citizenship Studies*, 14:4 (August 2010), pp. 381–94.
20. See Richard Boyd Barrett's account of the movement, 'Irish Water Protests', pp. 135–7.
21. Stephen Donnelly, 'Irish Water and a Raging Torrent of Inconvenient Truths', 7 April 2015, at: http://stephendonnelly.ie/irish-water-and-a-raging-torrent-of-inconvenient-truths/ (accessed 15 March 2015).

9. What is to be done?

1. Naomi Klein, *No Logo* (Toronto: Picador, 2000).

2. The Project for the New American Century (PNAC), founded by William Kristol and Robert Kagan, was a neo-conservative think-tank set up in Washington, D.C. in 1997 whose objective was 'to promote American global leadership'. Twenty-five people signed its founding statement, to which we refer here, ten of whom joined George Bush's presidential administration in key roles. They included Dick Cheney and Paul Wolfowitz.

3. It now operates with cleaner global products – health and education. The old US Steel building in the city centre is now the headquarters of a multinational health corporation.

4. Constance Elizabeth Hunt, *Thirsty Planet: Strategies for Sustainable Water Development* (London: Zed Books, 2004), p. 84.

5. Monsanto has teams of 'detectives' testing plant varieties and fining or threatening those who use varieties not paid for. In fact, farmers in the USA who do not use Monsanto products have been sued by the company because their patented seeds drifted onto neighbouring land and the offending farmers are then forced to stop using their own seed.

6. See, among many articles on the topic, Cormac Sheridan 'Doubts surround link between *Bt* cotton failure and farmer suicide', in *Nature Biotechnology*, 27 (2009), pp. 9–10. This reports on a study of whether the toxic pesticides were themselves responsible for the suicides. Clearly we are more convinced, following Vandana Shiva, that the debt burden is the main origin of this tragic phenomenon.

7. Maggie Black, *No-Nonsense Guide to Water* (London: Verso, 2004), p. 63.

8. Hunt, *Thirsty planet*, p. 63.

9. Ibid., p. 64.

10. Sandra Postel, *Pillar of Sand: Can the Irrigation Miracle Last?* (New York: Norton & Co., 1999).

11. Sandra Postel, 'Growing More Food with Less Water', *Scientific American*, 284: 2 (February 2001), pp. 46–50.

12. See the longer and more detailed account in Jonathan Neale, *Stop Global Warming: Change the World* (London: Bookmarks, 2008), pp. 233–39.

13. Juan Pablo Orrego S., 'La crisis mundial de las aguas: algunas de sus principales causas', Chile, March 2013, for the Fundación Terram, at: www.terram.cl (accessed 16 February 2015).

14. Naomi Klein, 'Reclaiming the Commons', *New Left Review*, 9 May–June 2001, pp. 81–89.

15. See Julian Caldecott, *Water: Life in Every Drop* (London: Virgin Books, 2007), pp. 118–19.

16. Annual report 2014 of Blacksmith Institute and Green Cross International, at: www.worldsworstpolluted.org (accessed 9 March 2015), p. 7.

17. Followed by Saudi Arabia $80.8bn at 9.3 per cent of GDP, Russia $70bn (4.1%), UK $61.8bn (2.3%), France $53.1bn (2.2%). All 2014 figures, Anup

Shah, 'World Military Spending', *Global Issues*, at: www.globalissues.org/article/75/world-military-spending

18. Annual report 2014 of Blacksmith Institute and Green Cross International, at: www.worldsworstpolluted.org, p. 16.

19. See the *UN World Water Development Report 2015*.

20. Margaret Thatcher's mantra about the 'nanny state', repeated by her successors, is one. See Satoko Kishimoto, Emanuele Lobina and Olivier Petitjean (eds), *Our Public Water Future: The Global Experience with Remunicipalisation* (London: PSRI, 2015).

21. There are an infinite number of examples, but perhaps South Africa can serve as representative, where while 10 million people had their electricity cut off in the decade after 1994, ANC officials in government were also receiving huge bribes from British arms companies. See Andrew Feinstein's *After the Party: Corruption, the ANC and South Africa's Uncertain Future* (London: Verso, 2009). The well-publicised case of the bribing of the Mayor of Grenoble by water companies is confirmation that the method is by no means limited to the developing world.

22. In Britain, 1.6 million people have an abnormally high concentration of nitrate in their blood in the wake of privatisation, for example; see Hunt, *Thirsty Planet*.

23. See Asit K. Biswas, 'Climate Change and Water Management', 16 October 2009, at: www.thirdworldcentre.org

24. Cf. Jessica Budds and Gordon McGranahan, 'Are the Debates on Water Privatization Missing the Point? Experiences from Africa, Asia and Latin America', in *Environment & Urbanization*, 15: 2 (October 2003), pp. 87–114.

25. Teflon, the non-stick material for skillets and saucepans, was apparently a by-product of the hugely expensive NASA space exploration programme. Useful as Teflon is, it is hard to believe it could not have been developed at much less cost.

26. Maude Barlow, '10 Water Commons Principles', *On the Commons*, 11 July 20 2012, at: www.onthecommons.org/work/10-water-commons-principles (accessed 2 April 2015).

27. David Harvey, *A Brief History of Neoliberalism* (Oxford: Oxford University Press, 2007), p. 19.

10. A new world water order

1. Asit K. Biswas, 'United Nations Water Conference Action Plan: Implementation over the past decade', *International Journal of Water Resources Development*, 4: 3 (September 1988), pp. 148–59.

2. The 2015 *UN World Water Development Report* does not make encouraging reading. Sixty-nine countries are described as 'seriously off course' for

completion of the sanitation goals, and 53 for drinking water. That means that they will be unlikely to achieve them even by 2030.

3. See Marcela Olivera's 'Water beyond the State: Letter from Cochabamba' and Richard Boyd Barrett's Report from the Irish Water Movement as examples.

4. See Satoko Kishimoto, Emanuele Lobina and Olivier Petitjean (eds), *Our Public Water Future: The Global Experience with Remunicipalisation* (London: PSRI, 2015).

5. Trevor Ngwane, 'Socialists, the Environment and Ecosocialism', *Monthly Review*, 20 November 2009.

Index